4th International Conference on Optics in Materials, Energy, and Technologies (ICOMET-2019)

IOP Conference Series: Materials Science and Engineering
Volume 720

Tainan, Taiwan
24 - 28 June 2019

ISBN: 978-1-7138-2579-1
ISSN: 1757-8981

Printed from e-media with permission by:

Curran Associates, Inc.
57 Morehouse Lane
Red Hook, NY 12571

Some format issues inherent in the e-media version may also appear in this print version.

Institute of Physics
Dirac House, Temple Back
Bristol BS1 6BE UK

Phone: 44 1 17 929 7481
Fax: 44 1 17 920 0979

techtracking@iop.org

Additional copies of this publication are available from:

Curran Associates, Inc.
57 Morehouse Lane
Red Hook, NY 12571 USA
Phone: 845-758-0400
Fax: 845-758-2633
Email: curran@proceedings.com
Web: www.proceedings.com

TABLE OF CONTENTS

IOP Conf. Series: Materials Science and Engineering **720** (2020) 011001 doi:10.1088/1757-899X/720/1/011001

Preface

We are proud to announce that the International Conference on Optics in Materials, Energy, and Technologies, 2019 (ICOMET-2019) will be held in National Cheng Kung, Tainan, Taiwan from June 24 to 28, 2019. Since its first conference in2015, this is the 4th one in the row. This conference brings together international leading experts, top scientists, young researchers, and students working about materials in Micro/Nano optics, Materials, Energy, and Technologies.

The ICOMET is a full-English conference . All the oral presentations are by invitation only and poster presentations are open for submissions. We would like to express our gratitude to all the invited speakers, the poster presenters, and attendees for your participation and involvement. The conference will not only provide a venue for in-depth scientific discussion but also an opportunity for finding potential new research partners, new forms of scientific collaborations, and also possible joint funding applications. Last but not the least, this conference is organized by "Light : Science and Applications" (LSA) Journal.

ICOMET-2019 conference serves as a great opportunity to establish collaborations and exchange ideas with colleagues in in Micro/Nano optics, Materials, Energy, and Technologies all over the world.

IOP Conf. Series: Materials Science and Engineering **720** (2020) 011002 doi:10.1088/1757-899X/720/1/011002

Peer review statement

All papers published in this volume of *IOP Conference Series: Materials Science and Engineering* have been peer reviewed through processes administered by the proceedings Editors. Reviews were conducted by expert referees to the professional and scientific standards expected of a proceedings journal published by IOP Publishing.

IOP Conf. Series: Materials Science and Engineering **720** (2020) 012001 doi:10.1088/1757-899X/720/1/012001

Recycle of vanadium from aluminum slag of ferrovanadium

Wei-Sheng Chen[1], Tzu-Yu Liu[2], Cheng-Han Lee[3]

[1, 2, 3] Department of Resources Engineering, National Cheng Kung University

[1] kenchen@mail.ncku.edu.tw

[2] j25227716@gmail.com

[3] happy980074@gmail.com

Abstract. Vanadium is a rare chemical element and it primarily comes from the by-product of other minerals. In this study, the material source mainly comes from vanadium-rich aluminum slag. This experiment aims to separate vanadium and other impurities, such as aluminum and zirconium. First, the slag was dissolved in the sodium hydroxide solution at pH value 13, and part of the alumina was precipitated. After removing the precipitation, the strong acid cation exchange resin, Dowex® G26, was added to eliminate the remaining impurities. In this procedure, the parameter of reaction time, the mass of ion exchange resin and pH value were supplied. Moreover, the adsorption isotherms described by means of the Langmuir and Freundlich isotherms were used to investigate the ion-exchange behaviors of vanadium. To sum up, this study will provide the best conditions of ion exchange to get high purity about 95% of vanadate.

Keywords: vanadium; ion exchange; aluminum slag; purify

1. Introduction

The compounds of vanadium have a wide range of applications such as steel industry, catalysts, automobiles, shipyard, fertilizers, ceramics, pigments and vanadium redox battery [1-4].85% of all vanadium is used to create alloys such as ferrovanadium. Ferrovanadium contains 35% to 80% vanadium.

It can obtain ferrovanadium by silicon reduction, and the process is used commercially [5-9]. In the process of producing ferrovanadium, scrap iron is melted, and a mixture of V_2O_5, aluminum and some flux is added. After the reaction, the aluminum is converted to alumina, and form the slag in the surface. The chemical composition of the slag varies depending on the raw materials, and it might contain some valuable metal, such as Vanadium and Gallium. The slag formed after producing ferrovanadium contained aluminum oxide (Al_2O_3), high percent of vanadium pentoxide (V_2O_5) which is used in the further processing of the metal and amount of Zirconium dioxide (ZrO_2). Ferrovanadium can be used in carbon steel, alloy steel high strength steel, and HSLA (High Strength Low Alloy) steel [10].

The first period of silicon reduction process is to restore vanadium from the fusion products of the third period of the previous heat. After the first period, the metal contains 25% to 30% V, 20% to 24% Si, and 0.3% to 0.5% C and V_2O_5 content 0.25% to 0.35% in the slag. In the second period, the operations are to restore additional V_2O_5 with silicon and aluminum to increase vanadium content in the metal. At this period, the slag contains 8% to 10% V_2O_5.The second period ends when the metal composition reaches 35% to 40% V, 9% to12% Si, and 0.4% to 0.6% C. The V_2O_5 content in the slag drops to<0.35%V_2O_5.In the third period more lime and vanadium pent-oxide are added, and the purpose is to utilize reaming silicon in the metal to reduce more vanadium. The slag from the third period contains 12% to 16% V_2O_5.Ferrovanadium can be also prepared by the thermite reaction. In the process, vanadium and iron oxides are co-reduced by aluminum granules [8-9].

Sodium carbonate roasting technology is used to extract vanadium from the vanadium-bearing slag. In this process, Na_2CO_3 is used as a calcining additive, and $MgSO_4$ is added as a conversion agent. By using this process, vanadium could be extracted, but leaching rate is low, only about 65% to 75%, and the aluminum can't be recovered effectively [11].

Finding a way to separate the vanadium from the vanadium-rich slag is the main goal in this experiment. After leaching the slag with NaOH solution, it can obtain vanadium-contained solution. Compared with the methods mentioned above, the ion exchange method has the advantages of, such as high purification efficiency, less pollution, short reaction time, and the reusability of the resin. In the purification process, the cation ion-exchange resin is chosen to remove the impurities. Dowex G26, a strongly acidic cation exchange resin, has a great adsorption efficiency on the impurities, aluminum and zirconium. Because the colorless vanadate VO^{3-} ions in alkaline medium. To sum up, the leaching method and the Dowex G26 resin were used to get a high purity of the vanadium.

2. Materials and experiments

2.1. Materials

The aluminum slag which is used in this study comes from the production of ferrovanadium. The slag is composed of 49.67% Al_2O_3, 49.31% V_2O_5 and 0.82% ZrO_2. The content of the slag detected by XRF analysis is shown in Table.1.

Table 1. The content of the slag

Element	V_2O_5	Al_2O_3	ZrO_2
Concentration (%)	49.31%	49.67%	0.82%

Sodium hydroxide (NaOH) was purchased from Sigma-Aldrich (St. Louis, MO, USA) (NaOH ≥ 98%), for the leaching step. Hydrogen chloride (HCl) and sulfuric acid (H_2SO_4) were purchased from Sigma-Aldrich (St. Louis, MO, USA) (HCl ≥ 36.5%)(H_2SO_4 ≥ 98%) to adjust the pH value. Dowex G26 was obtained from Sigma-Aldrich (St. Louis, MO, USA) and was used as a strong acidic cation exchange resin, to remove the impurities. Multi-elements ICP standard solutions were acquired from AccuStandard, New Haven, Connecticut State, USA. The nitric acid (HNO_3) were acquired from Sigma-Aldrich (St. Louis, MO, USA) (HNO_3 ≥ 65%). Ammonium hydroxide (NH_4OH) and Ammonium chloride (NH_4Cl) were obtained from Honeywell and Sigma-Aldrich (St. Louis, MO, USA) respectively (NH_4OH, 30-33%) (NH_4Cl ≥ 95%). (Table.2 and Table.3 show the information of the materials)

Table 2. The information of the materials

Chemical Composition	Purity	Lable
NaOH	98% ≥	Applichem Panreac
HCl	≥ 36.5%	Sigma-Aldrich
HNO_3	65% ≥	Sigma-Aldrich
NH_4OH	30-33%	Honeywell
H_2SO_4	98% ≥	Sigma-Aldrich
NH_4Cl	99.5% ≥	Sigma-Aldrich

ICP multi-element standard solution IV	---	Merck

Table 3. The information of Dowex G26

	Dowex G26
Physical form	Brown opaque spherical beads
Matrix	styrene-divinylbenzene(gel)
pH group	0-14
Function groups	Sulfuric acid
Moisture holding capacity	45-52%
Particle size	0.6-0.7mm
Total exchange capacity	2.0 meq/mL

2.2. Equipment

The materials were analyzed by inductively coupled plasma optical emission spectrometry (ICP-OES; Varian, Vista-MPX, PerkinElmer, Waltham, MA, USA).

2.3. Experimental Procedures

2.3.1. Leaching

This process is aims to leaching the vanadium from the slag. Figure 1 is the Eh–pH diagram of vanadium. According to the figure, vanadium pentoxide would turn into the colorless vanadate VO^{3-} ions in alkaline solution (The reaction is shown in Eq. 1).

$$V_2O_5 + NaOH \rightarrow Na_3VO_4 + H_2O \qquad \text{Eq. 1}$$

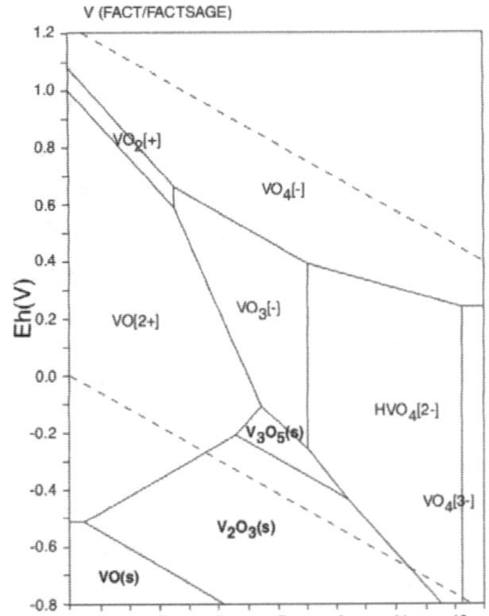

IOP Conf. Series: Materials Science and Engineering **720** (2020) 012001 doi:10.1088/1757-899X/720/1/012001

Figure 1. The Eh–pH diagram of vanadium.

2.3.2. Ion-Exchange

Dowex G26 is a strongly acidic resin which was used to exchange cations ions efficiently. In this study, we use Dowex G26 to remove the Aluminum and Zirconium ion which are the most two impurities in the solution. First, we use the adsorption isotherms described by means of the Langmuir and Freundlich isotherms to investigate the ion-exchange behaviors of Aluminum ion. Moreover, we compare the Aluminum and Zirconium ion adsorption efficiency at pH value from 12.5 to 14 to find the best parameter. With the parameters of pH value, we also compare the efficiency with the mass of ion exchange resin and reaction time (from 2 minutes to 1024 minutes). After removing the impurities, the content of the vanadium solutions was investigated in this study. The whole experiment process is shown in figure.2.

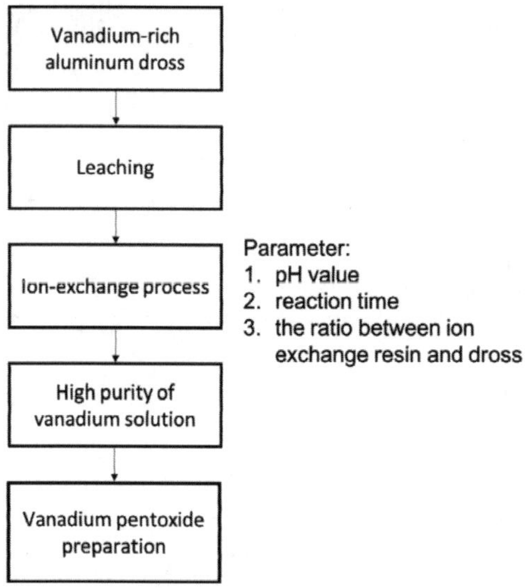

Figure 2. The outline of experiment process

3. Result

3.1. Leaching

In this process, the 0.2 grams slag was dissolved in 100 milliliter Sodium Hydroxide solution, and part of the aluminium oxide (Al_2O_3) would be residue as the solid phase. In the experience, the lower the pH value is, the more the aluminium oxide (Al_2O_3), would be found at the bottom. As Figure 3 shows that the leaching of vanadium is 98.1%, the leaching of aluminium is 8.66% and the leaching of Zirconium is about 28.51% with the pH value is 12.5. At this condition, the maximum residues of aluminium oxide (Al_2O_3) is 0.0785g.

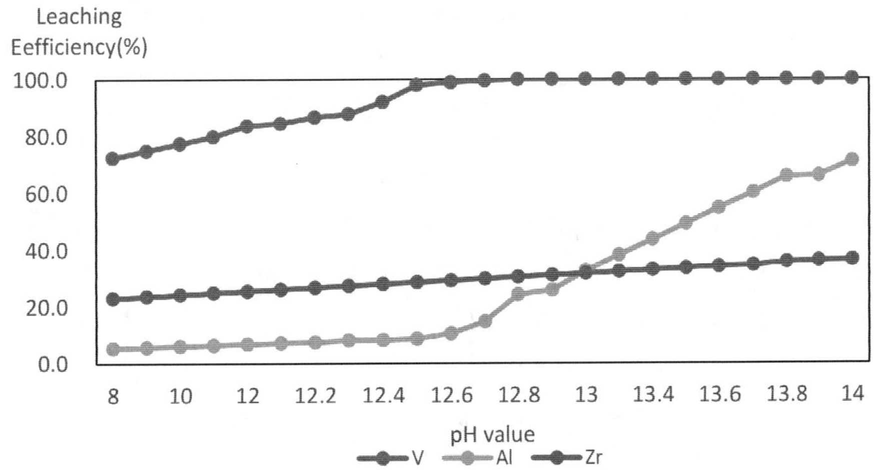

Figure 3. Leaching efficiency on V, Al, Zr with pH value

3.2. Ion-Exchange

3.2.1. Isothermal Adsorption Models

In this experiment, the varies initial concentrations of aluminium (10,20,50,100,200,300 to 500 ppm) and the relationship between $C_e$and q_e were used to create the isothermal adsorption curve. (C_e: the equilibrium concentration of adsorbate (mg/Lc), q_e: the amount of metal adsorbed per gram of the adsorbent at equilibrium (mg/g)) Figure 4 shows the isothermal adsorption curve of aluminum. The saturated adsorption amount was about 208 milligram per gram.

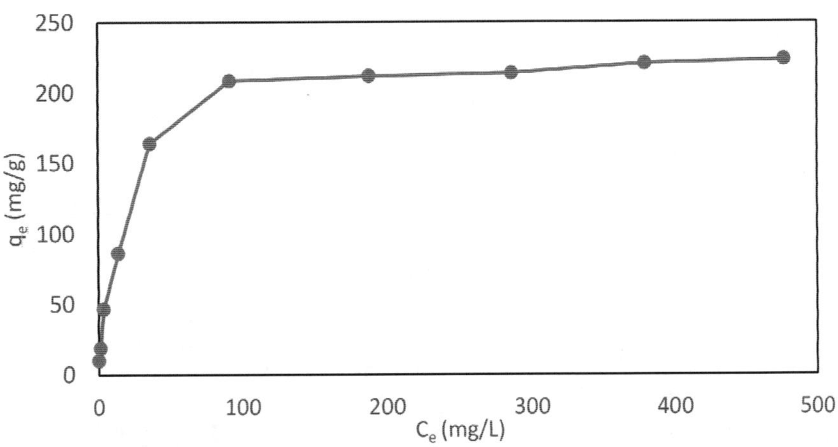

Figure 4. Isothermal adsorption curve of aluminum

To get the adsorption behavior of Dowex G26, Langmuir and Freundlich isothermal adsorption model were used and the regression coefficients were compared. Equation 2 show the Langmuir equation [11]. Figure 5 is the Langmuir isothermal adsorption curve. Equation 3 show the Freundlich equation [11]. Figure 6 is the Freundlich isothermal adsorption curve. By comparing the R square in figure 5 and 6, the adsorption behavior of Dowex G26 more fits with the Langmuir adsorption model.

IOP Conf. Series: Materials Science and Engineering **720** (2020) 012001 doi:10.1088/1757-899X/720/1/012001

It means that the adsorption of the Dowex G26 is uniform and no transmigration of adsorbate in the plane of the surface.

$$\frac{C_e}{q_e} = \frac{C_e}{q_m} + \frac{1}{q_m K_L} \quad \text{Eq. 2}$$

Where:

C_e: the equilibrium concentration of adsorbate (mg/L)

q_e: the amount of metal adsorbed per gram of the adsorbent at equilibrium (mg/g)

q_m: maximum monolayer coverage capacity (mg/g)

K_L: Langmuir isotherm constant (L/mg)

$$lnq_e = lnK_F + \frac{1}{n} lnC_e \quad \text{Eq. 3}$$

Where:

C_e: the equilibrium concentration of adsorbate (mg/L-1)

q_e: the amount of metal adsorbed per gram of the adsorbent at equilibrium (mg/g)

n: adsorption intensity

K_F: Freundlich isotherm constant (mg/g)

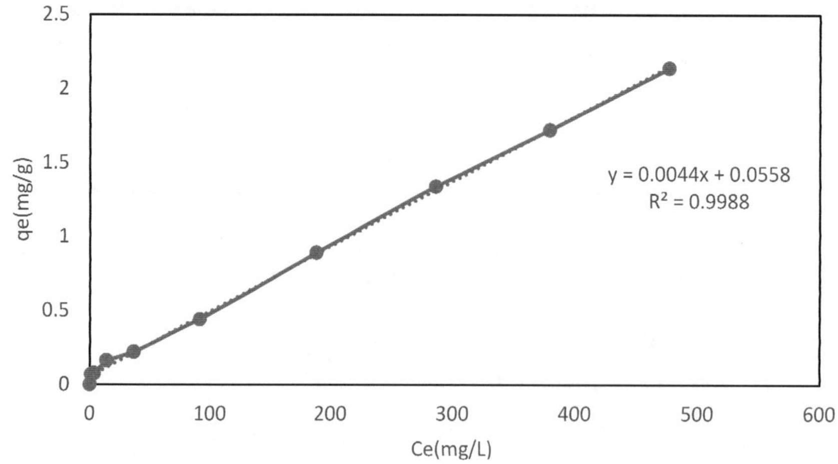

Figure 5. Langmuir isothermal adsorption curve

IOP Conf. Series: Materials Science and Engineering **720** (2020) 012001 doi:10.1088/1757-899X/720/1/012001

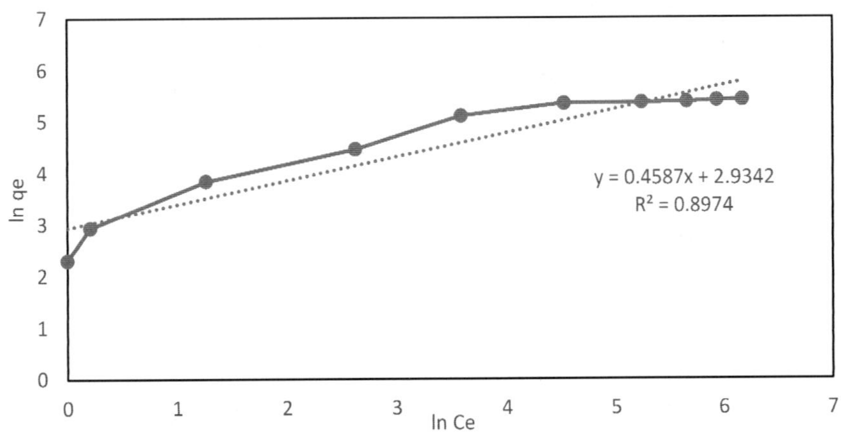

Figure 6. Freundlich isothermal adsorption curve

3.2.2. Effect of pH value

The solution after leaching is used to investigate the effect of pH value with a range from 12.5 to 14. Figure 7 shows the relationship between the pH value and the adsorption efficiency of Aluminum and Zirconium ion. As seen in the figure, pH value 13 has the greatest adsorption efficiency of Zirconium ion, about 99.89%. The adsorption efficiency of Aluminum has the best parameter at pH 12.5 (90.94%) and the adsorption efficiency is 89.59% at pH 13. Due to this condition, pH 13 was chosen to be the best parameter.

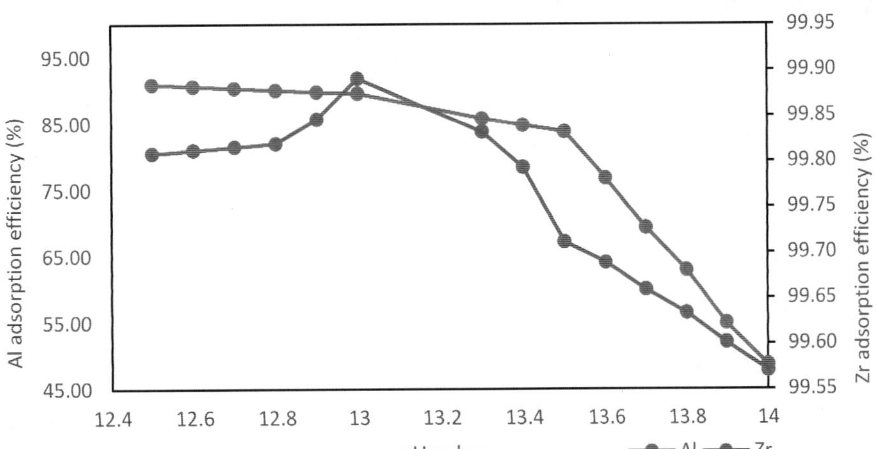

Figure 7. The relationship between the pH value and the adsorption efficiency of Al and Zr

3.2.3. Effect of Reaction Time

The reaction time was investigated from 2 to 1024 minute. Figure 8 shows the result of adsorption efficiency. In the figure, at 4 minutes the adsorption efficiency on zirconium was 100%. The adsorption efficiencies on aluminum from 2 minutes to 1024 minutes were almost the same (89.54%-92.95%). 4 minutes was chosen to be the best reaction time parameter.

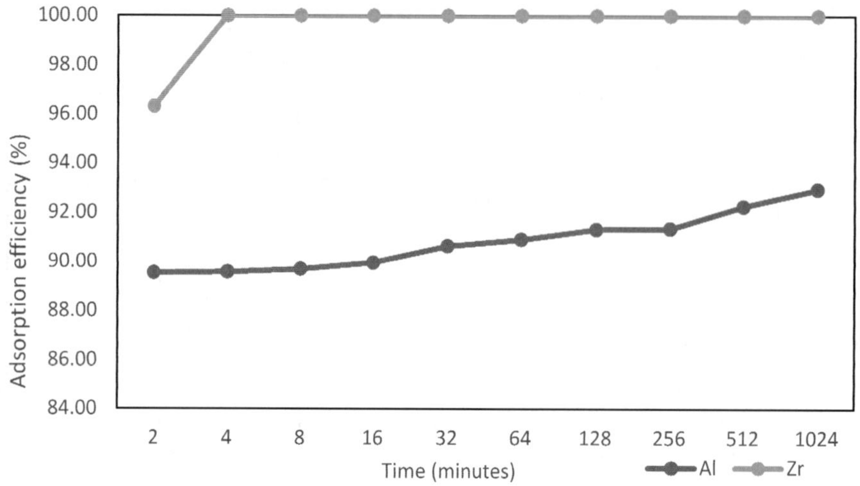

Figure 8. The adsorption efficiency with reaction time

3.2.4. Effect of the mass of ion exchange resin

Figure 9 shows the adsorption efficiency with the mass of ion exchange resin from 0.1g to 1.5g. According to this graph, when the mass of resin was 0.2g. The adsorption efficiency of Zr approached 100%. With the mass of resin was 1.2 g and 1.5 g, the adsorption efficiencies of Al were almost the same (90.84% and 92.58% respectively). Due to this condition, 1.2 g resin was chosen to be the best parameter.

After getting the optimum parameters of reaction time, the mass of ion exchange resin and pH value, the vanadium solution were purified, and the content detected by ICP is shown in table 4.

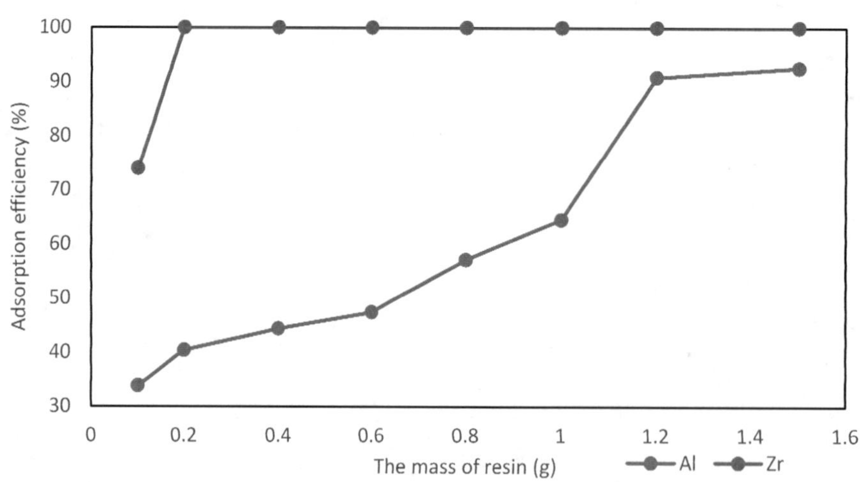

Figure 9. The adsorption efficiency with the mass of ion exchange resin

Table 4. The content of vanadium solution

Element	V	Al	Zr
Content (%)	99.76%	0.24%	N.D.

4. Conclusion

1. The leaching efficiency of vanadium is 98.1% at the optimum condition of pH 12.5.

2. pH value 13.5, 4 minutes reaction time, 1.2g resin were the optimum parameter of Dowex G26 in ion-exchange step to purify the impurity.

3. The purity and recovery rate of vanadium product is 99.758% and 96.83%.

4. The process in this study can provide a method to purify vanadium effectively.

5. Acknowledgments

This study was supported by National Cheng-Kung University Dept. of Resource Engineering, R.O.C.

References

[1] Imtiaz M, Rizwan M, Xiong S et al. Vanadium, recent advancements and research prospects: A review. Environ Int. 2015;80:79-88. doi:10.1016/j.envint.2015.03.018
[2] B. Rohrmann, "Vanadium in South Africa," J. S. Afr. Inst. Min. Metall. 85(5), 141–150 (1985).
[3] Baroch, E.F.,. Vanadium and vanadium alloys. In: Kirk-Othmer Encyclopedia of ChemicalTechnology, volume 24. John Wiley & Sons, New York, pp. 782–797. 2005
[4] Lourenssen K, Williams J, Ahmadpour F, Clemmer R, Tasnim S. Vanadium redox flow batteries: A comprehensive review. Journal of Energy Storage.;25:100844. doi:10.1016/j.est.2019.100844,2019
[5] Swinbourne D, Richardson T, Cabalteja F. Understanding ferrovanadium smelting through computational thermodynamics modelling. Mineral Processing and Extractive Metallurgy. 2016;125(1):45-55. doi:10.1179/1743285515y.0000000019
[6] Vermaak , M. K. G. Vanadium recovery in the electro-aluminothermic production of ferrovanadium. thesis, 2000
[7] Human Eric R. Treatise On Process Metallurgy, Volume 3: Industrial Processes. Elsevier Science; 2013.
[8] Gasik, M.I., Lyakishev, N.P., Gasik, M.M., 2009. Physical Chemistry and Technology of Ferroalloys [in Ukrainian]. Dnipropetrovsk: Sistemnye Tehnologii, 494 pp
[9] Gasik M. Handbook Of Ferroalloys. Oxford: Elsevier Science; 2013.
[10] Tian, P.; Zhong, Z.; Bai, R.; Zhang, X.; Gao, H. Application of Different Vanadium Alloys in Steel. Proceedings of the International Conference on Computer Information Systems and Industrial Applications 2015, 861-864.
[11] GAO Ming-lei CHEN Dong-hui SHI Li-xin LI Lan-jie ZHOU Bing-jing QI Jian. Recovery of Vanadium and Aluminum in Ferrovanadium Slag from Thermite Method by Sodium Carbonate-roasting. , 2012, 12(4): 576-582
[12] A.O D. Langmuir, Freundlich, Temkin and Dubinin–Radushkevich Isotherms Studies of Equilibrium Sorption of Zn 2+ Unto Phosphoric Acid Modified Rice Husk. IOSR Journal of Applied Chemistry. 2012;3(1):38-45. doi:10.9790/5736-0313845

Using Synergy Approach To investigate the developmental trends of Solar Energy

Yi-Huei Su [1, a], a and Jung-Hua Wu [2,b*]

[1]Department of Resources Engineering, National Cheng Kung University, Tainan 701, Taiwan

n48021050@ncku.edu.tw

[2]Department of Resources Engineering, National Cheng Kung University, Tainan 701, Taiwan

hwaa@mail.ncku.edu.tw

*Corresponding author: hwaa@mail.ncku.edu.tw

Keywords: Synergy Theory; Solar Energy; Patent; Journal Paper; Research and development budget

Abstract. In recent years, electricity generation using solar energy has become a popular trend. At present, the solar energy output value of China, the United States (US), and the European Union (EU28) together account for more than 70% of the total global output value; therefore, these nations have a significant influence on future development of solar energy technologies.

In this study, we analyzed data, including those for patents; key technology; output value; trade surpluses (deficits); research and development (R&D) funds and manpower; energy density, and publications in journals, and collected data related to solar energy use in these countries to analyze trends in its use by applying the synergy theory. The subsystems investigated include the industry, government, and academic, and the study period was 2005–2015. We also integrated resources and linked application statuses to track applications by the considered subsystems.

The following results were obtained at the end of this study:

(1) Development of solar energy use technology has slowed down in the EU28, is growing in China, and has plateaued in the US. In terms of trends in synergy, the EU28 showed growth up to 2013, after which a rapid decline was observed; an overall growth trend was observed in China; and a plateau in the was noted in the US after 2011.

(2) Overall, variations in these trends reduced across the industry subsystems affecting decision-making in companies. Overall investment in research and development (R&D) by the government has plateaued. There appeared to be a growth in the academic subsystems, with the number of studies on solar energy increasing.

(3) Patents and published journal articles indicate that, in the EU28, research on key technology for solar energy use has been dominated by theory. However, in China and the US, research is mainly focused on the application of such technology.

The observed trends for solar energy technologies in the study areas. However, the environment for the solar energy use can change; it is recommended that future research take market considerations into account to better understand the trends in the development of solar energy technologies.

1. Introduction

In recent years, issues regarding solar power generation have gradually become the focus of global research attention. The current total solar energy output value of China, the United States (US), and the European Union (EU28) together constitute more than 70% of the total global output value[1-2]. To enhance the competitive power, the industry applies for patents for development results, the government invests in research funding and manpower, and the academic community publishes research results in journals. We also integrated resources and linked application statuses to track applications by the considered subsystems, and collected data related to solar energy use in these countries to analyze trends.

This study focuses on China, the United States, and the European Union as research areas, based on the synergy theory approach. The study period is from 2005 to 2015. The research scope is limited to an analysis of the subsystems of industry, government, and the academia, and data were collected from the World Intellectual Property Organization (WIPO), Organisation for Economic Co-operation and Development (OECD), and Web of Science®, respectively.

2. Technologies Utilized in Solar Energy

The development of clean energy sources became a global trend after the Second Industrial Revolution. Improvements in the energy conversion efficiency can produce benefits such as lowered greenhouse gas emissions, safe energy, and an improved energy supply [3]. The most important issues related to solar energy are the low efficiency of conversions and energy storage capacity. However, the methods currently available to resolve these issues are insufficient [4]; the most popular key solar technologies involve techniques such as "light concentration," "Förster energy transfer," "conductive materials," "plasmons," "nanoantennas," and "quantum energy transport." Research on the key solar energy technologies have tended towards the goal of improving the efficiency of solar energy [5].

Light concentration improves the focus density of solar cells and can result in increases in the conversion efficiency per unit area [6-8]. The plasmonic effect plays a very important role in increasing local solar power density and thus in efficient solar cell conversion [9-10].

Traditional solar cells absorb visible light during the day but not at night; however, nanoantennas can absorb radiation energy from mid-infrared light and can thus be used both during the day and at night, when their surroundings continue to radiate infrared light. Thus, they can improve the efficiency of solar energy conversion by prolonging the time for which devices utilizing solar power can be used [11].

Manipulating the resistance of solar cells can also improve conversion efficiency. Transparent conductive films are low-resistance materials that generally improve the conductivity and transparency of the devices; however, they need to be very thin [12].

In 1982, Professor Arakawa Taehoon, director of the Nano-Quantum Information Electronics Research Organization at the University of Tokyo, pioneered the concept of quantum dots. The current conversion efficiency for solar cells using polycrystalline silicon is about 15%; however, those employing monocrystalline silicon offer a conversion efficiency of 20%. Although the use of crystalline silicon has increased efficiency, the latter is still less than the level desired. Quantum and coherent energy transfer methods, e.g., Förster energy transfer, are key technologies expected to lead to increase the absorption of solar energy [13-20].

These studies suggest that the use of light concentration, nanoantennas, conductive materials, quantum dot technologies, plasmons, and Förster energy transfer methods can improve solar energy absorption. They also suggest that the future direction for research should be based on such considerations.

Other paragraphs are indented (BodytextIndented style).

3. Basic Equations

Our approach uses the synergy theory proposed by the German physicist Hermann Haken in 1974. [21-22] We combined statistics with dynamics to integrate applications between systems and determine the causal relationships among factors with the aim to explore the interactions between industry, government, and the academia.

The theory can be expressed as the following basic formulas:

$$\frac{du}{dt} = -KU + g(I, G, A) + F \quad \text{.....(1)}$$

$$\frac{du}{dt} = -K_1U + g_1(I, G, A) + F_1 \quad \text{.....(2)}$$

$$\frac{du}{dt} = -K_2U + g_2(I, G, A) + F_2 \quad \text{.....(3)}$$

$$\frac{du}{dt} = -K_3U + g_3(I, G, A) + F_3 \quad \text{.....(4)}$$

where

t: time;

K: coefficient describing the relationship between the rate of change and the original state;

K_1, K_2, K_3: coefficients describing the relationships between the rates of change in I, G, and A, respectively and the original state;

U: uncorrected degree of synergy;

g: effect of the subsystem synergy on the development of collaborative systems;

g_1, g_2, g_3: impact of the synergy of the three subsystems on E, R, and P, respectively;

I : industry subsystem;

G: government subsystem;

A: academia subsystem;

F: constant external force;

F_1, F_2, F_3: constant external force in individual subsystems;

According to Equation (1), the synergy between the application status trajectories of the three subsystems I (industry subsystem), G (government subsystem), and A (academic subsystem) show that they can be integrated. By analogy, Equations (2), (3), and (4) represent the synergy between the application trajectories of the EU28, China, and the US over the years.

Let S = $\{S_1, S_2, \ldots S_j \ldots S_n\}$ and j = 1, 2, …, n, where the set of order parameters in the subsystem is S_j. $e_j=\{e_{j1},\ldots,e_{ji},\ldots,e_{jm}\}$, $i \in [1,m]$; $e_{ij} \in [\beta_{ji},\alpha_{ji}]$.

α_{ji}: the limit value of the system order parameter e_{ji};

β_{ji}: the limit value of the system order parameter e_{ji}.

The order degree u_j can be determined as follows:

$$u_j(e_j) = \begin{cases} (e_{ji}-\beta_{ji})/(\alpha_{ji}-\beta_{ji}) \\ (\alpha_{ji}-e_{ji})/(\alpha_{ji}-\beta_{ji}) \end{cases} \quad \text{.....(5)}$$

where $u_j(e_j) \in [0, 1]$.

$$u_j(e_{ji}) = \sum_{i=1}^{m} \lambda_i u_j(e_{ji}),\ \lambda \geq 0,\ \sum_{i=1}^{m} \lambda_i = 1 \quad \text{.....(6)}$$

As shown in Equation (6), the following model formula is used:

$u_j(e_{ji})$ is the order degree of subsystem S_j, given $u_j(e_{ji}) \in [0,1]$

λ_i is the weight of e_{ji}

$u_j^0(e_{ji})$ is the subsystem order value, for example t_1, expressed as $u_j^1(e_{ji})$. Then,

$$U = \theta \ X \ \sqrt[n]{\prod_{i=1}^{n} \left| u_j^1(e_j) - u_j^0(e_j) \right|} \ldots\ldots\ldots\ldots\ldots\ldots\ldots\ldots\ldots\ldots\ldots\ldots\ldots (7)$$

where

$$\theta = \frac{\min_j [u_j^1(e_j) - u_j^0(e_j)]}{\left| \min_j [u_j^1(e_j) - u_j^0(e_j)] \right|}, \text{ among them, } u_j^1(e_j) - u_j^0(e_j) \neq 0 \ldots\ldots\ldots\ldots\ldots (8)$$

4. Developments in Solar Energy Technology in Industry, Government, and Academia

This study uses the World Intellectual Property Organization (WIPO) as the basis for an investigation data of solar cell patents. All journals can be obtained from the Web of Science® database. Research and development funding sources were determined from data taken from the Organization for Economic Co-operation and Development (OECD) database on research funding statistics.

4.1 The European Union (EU28)

The number of solar energy patent applications remained generally stable in the EU28 from 2005 to 2008 (see Figure 1). The number of the approved solar patents show that the top three technologies in this market involved the use of "light concentration," "plasmons", and "conductive materials." The top three research fields, based on the number of papers published on solar energy, are "conductive materials," "light concentration," and "nanoantennas."

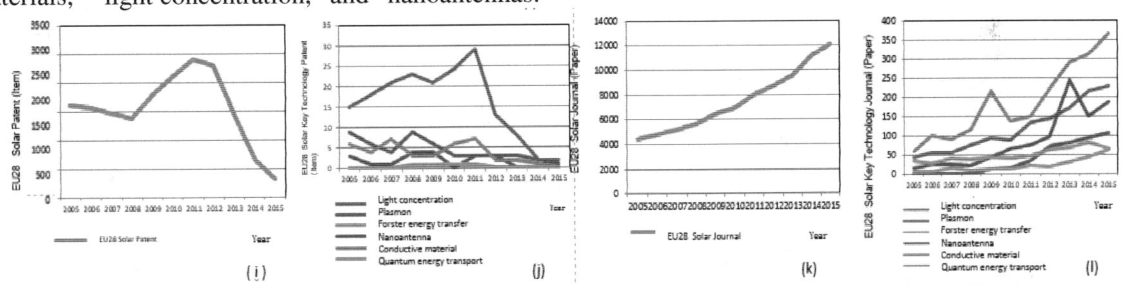

Figure 1. Trends in the number of (a) patents and (b) key solar technology patents filed, (c) journal articles published and (d) key journal articles, related to solar technology, for the EU28 countries from 2005 to 2015.

4.2 China

From 2005 to 2015, the Chinese government invested a lot of money on scientific research, patents, and journal papers, as shown in Figure 2. Based on the number of approved solar patents, the top three technologies of the interest during this time period were "nanoantennas," "conductive materials," and "quantum energy transport." The top three research fields, based on the number of papers published on solar technology, were "conductive materials,"

IOP Conf. Series: Materials Science and Engineering **720** (2020) 012003 doi:10.1088/1757-899X/720/1/012003

"nanoantennas," and "light concentration."

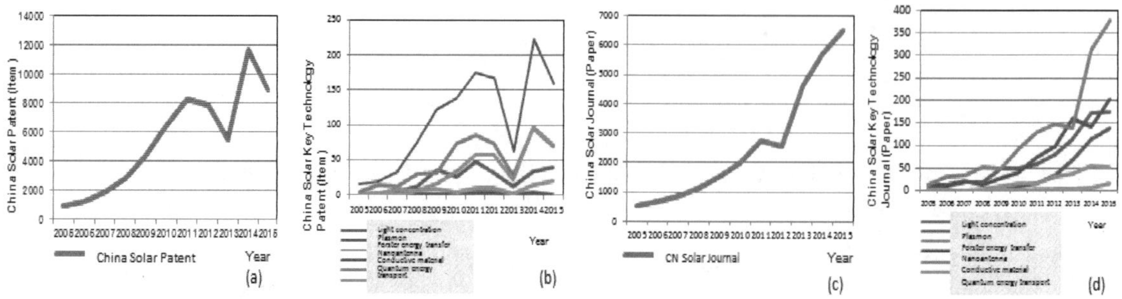

Figure 2. Trends in the number of (a) patents and (b) key solar technology patents filed, (c) journal articles published and (d) key journal articles, related to solar technology in China from 2005 to 2015.

4.3 The United States

In the US, the number of patents related to solar technology first showed an increasing trend and then began trending downwards in 2012; this had an indirect relationship with the input for research funding, as shown in Figure 3. Based on the number of the approved patents for solar technology, the top three technologies under investigation were "conductive materials," "nanoantennas," and "quantum energy transport." The top three research fields, based on the number of papers published on published solar technology, are "conductive materials," "light concentration," and "nanoantennas."

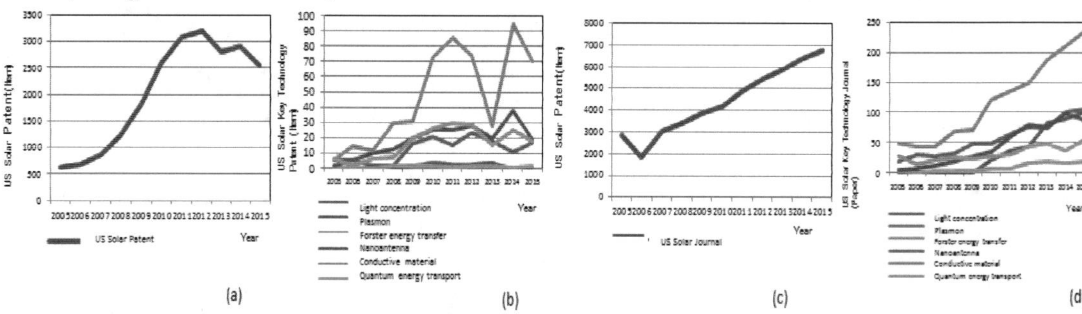

Figure 3. Trends in the number of (a) patents and (b) key solar technology patents filed, (c) journal articles published and (d) key journal articles, related to solar technology in the United States from 2005 to 2015.

5. Results and Discussion

This study attempted to integrate and present data on the current status of the solar energy technology industry, related government policy, and academia, and to speculate on future trends and provide strategic advice, using big data operations. Until 2011, the synergy between the EU28, China, and the US showed steady growth, with the EU28 ahead of the other two countries (Figure 4). In 2013, there was a divergence in this trend; discovery continued to grow in China, plateaued in the US, and declined in the EU28.

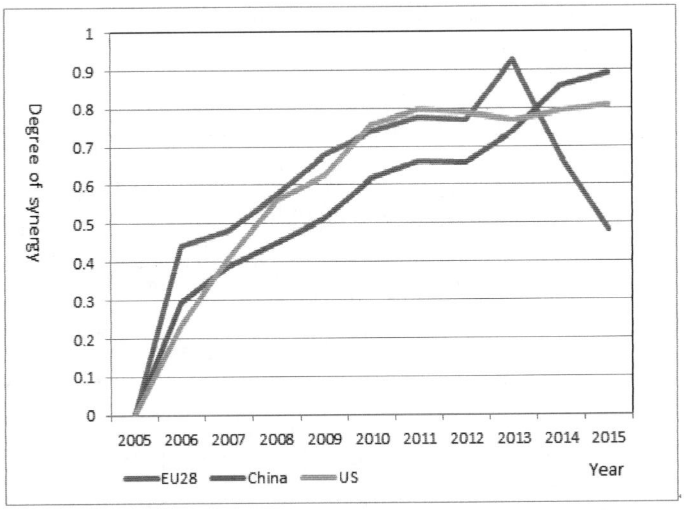

Figure 4. Degree of synergy in patterns of interest in solar energy technology in the EU28, China, and the US.

The trends in the subsystems of industry, government, and academia can be described as follows (see Figure 5):

For industry, the EU28 had a steady growth in the period 2005–2009 but began to decline sharply after a turning point in 2013 (Figure 5(a)). The reasons for this could be a sharp decrease in solar output (the 2012 value was half of the preceding year); a sharp decrease in the trade surplus (deficit) in 2015, when it was one quarter of that in 2011; and a decrease in the number of patents issued in the same year, which was one-eighth that in 2011. In 2009, 2012, and 2014, China experienced a decline in market volatility owing to a slight decline in output and a trade surplus (deficit). The US market enjoyed growth in 2009 before experiencing a shock from 2009 to 2013 and then a slowdown owing to a decrease in the number of patents approved.

The patents, output, and trade surplus (deficits) invested in by the industry over the years responded to changes in the industrial environment of a particular country. When the EU28 suffered an economic decline; the US economy declined steadily as well and when the Chinese market increased and then slowed down, a corresponding diminishing trend was noted in the industrial environment. Companies need to make steady decisions carefully considering the trends in industry, government, and academia.

In terms of government subsystems, the EU28 experienced growth during the period 2005–2012, after which it gradually declined (Figure 5(b)) because the budget for government R&D expenditures decreased annually after 2012. 2005–2012 marked a period of rapid growth in China followed by a plateau after 2013, the latter again owing to a levelling-out in the budget for government R&D expenditure and government R&D manpower. The US experienced a shock during the period 2007–2011, with the budget for government R&D expenditure changing only slightly annually 2007 to 2011, before stabilizing in 2012 and remaining level since then.

The R&D expenditure and manpower that governments have invested in solar energy technology over the years has reflected the state of investment in R&D in the country. Overall trends, and therefore the intensity of government investments in R&D on solar energy technology, were flat for the period of interest for this study.

In terms of the academia subsystems, the EU28 enjoyed growth from 2005 to 2013 and a decreasing trend after 2014, as shown by the slight decline in the number of journal publications after that time. However, there was a rapid growth in China from 2012 to 2014 owing to a spike in the number of journal articles published. From 2005 to 2013, there was growth in the US; however, this gradually reached a plateau after 2014 as the trend in the number of journal articles published after 2014 flattened.

The number of journal articles published over the years reflects the state of research in a country; as the overall trend increases, there is an implied increase in research output.

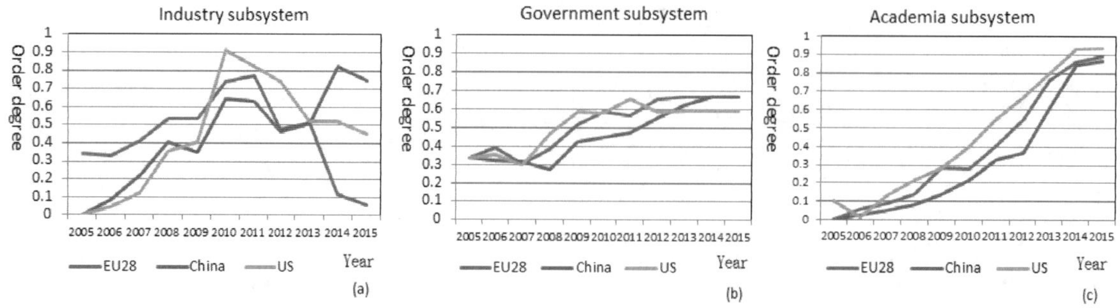

Figure 5 Trends noted for the order of the degrees of synergy for (a) industry, (b) government, and (c) academia subsystems, related to solar energy technology in the EU28, US, and China.

The trends in the key technologies related to solar energy discussed in section 4.1–4.3, applicable to solar-powered devices and classified by country (Figure 6), show that developments related to solar-based thermal technology was better in the EU28, whereas China and the US fared better in terms of solar photovoltaic technology. The key technologies related to solar energy will affect the trend in development of solar energy technologies in these countries.

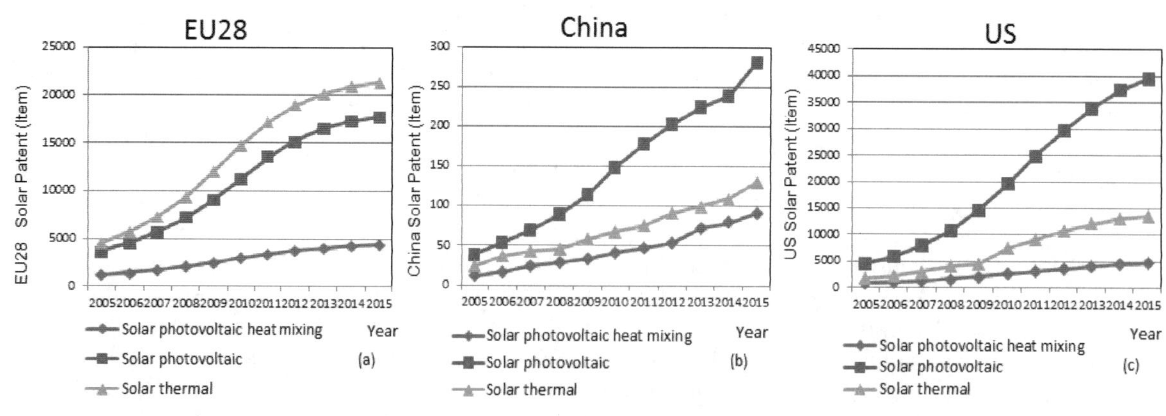

Figure 6. Trends related to solar photovoltaic heat mixing, general solar photovoltaics, and solar thermal technologies in (a) the EU28, (b) China, and (c) the United States. [24]

6. Conclusion

In summary, the results of this study show that a rise in the development or improvement of key technologies related to solar energy does affect the development of devices and research in the area. The EU28 dominate the field of theoretical studies, in terms of patents filed and journal articles published whereas China and the US are engaged mainly in applying key technologies.

Our research also showed that:

(1)In terms of trends in synergy, the EU28 showed growth up to 2013, after which there was a rapid drop. This could be attributed to the fact that, solar energy output value; patents; and import and export values for the region fell sharply after 2013. An overall growth trend was seen in China; after 2011, there was a plateau in the trend for the US, due mainly to small fluctuations in solar energy output value, patents, and a trade surplus (deficit).

(2)The degree of the change in the trends was, overall, downward for the industry subsystems, although the EU28, China, and the US showed diverging trends in 2013; the former did however affect the decisions taken by companies. For the government subsystem, overall trends were flat; therefore, trends for investment in R&D were flat. The academia subsystem exhibited a growing trend, as the annual production of journal articles increased and, as a result, the number of studies undertaken on solar energy also increased.

This study shows that, in terms of trends for solar energy technologies, the EU28 is in recession, China is growing, and the US has plateaued. In the industry subsystems, there has been a continuous downward trend continues since the divergence described in this paper. This may affect corporate decision-making; however, the environment for the solar energy industry can change, it is recommended that future research take market considerations into account, so that it can better understand the trends in the development of solar energy.

References

[1]Ministry of Economic Affairs International Trade Bureau,Taiwan.

[2]Bureau of Foreign Trade in Ministry of Economic Affairs,Taiwan.

[3]International Energy Agency 2015 Capturing the Multiple Benefits of Energy Efficiency

[4]Castillo C.P, Silva F.B.E and Lavalle C 2016 An assessment of the regional potential for solar power generation in EU-28,Energy Policy. 88 86-99.

[5]Guo C.F, Sun T.Y, Cao F, Liu Q and Ren Z.F, 2014 Metallic nanostructures for light trapping in energy-harvesting devices, *Light: Science & Applications.* 3 e161.

[6]Memarian M and Eleftheriades G.V 2013 Light concentration using hetero-junctions of anisotropic lowpermittivity metamaterials, *Light: Science & Applications.* 2 e114.

[7]Kosten E.D, Atwater J.H, Parsons J, Polman A and Atwater H.A 2013 Highly efficient GaAs solar cells by limiting light emission angle, *Science & Applications.* 2 e45.

[8]Holman Z.C, De W.S and Ballif C 2013 Improving metal reflectors by suppressing surface plasmon polaritons: a priori calculation of the internal reflectance of a solar cell, Light: Science & Applications. 2 e106.

[9]Guo C.F, Sun T.Y, Cao F, Liu Q and Ren Z.F 2014 Metallic nanostructures for light trapping inenergy-harvesting devices, *Light: Science & Applications.* 3 e161.

[10]Chen X, Jia B.H, Zhang Y.A and Gu M 2013 Exceeding the limit of plasmonic light trapping in textured screen-printed solar cells using Al nanoparticles and wrinkle-like graphene sheets, *Light: Science & Applications.* 2 e92.

[11]Taylor R.A, Otanicar T, Rosengarten G 2012 Nanofluid-based optical filter optimization for PV/Tsystems, *Light: Science & Applications.* 1 e34.

[12]Katsuhiro P. I and Kazushi M 2014 A visible light-driven plasmonic photocatalyst, Francesca Pincella, *Light: Science & Applications.* 3 e133.

[13]Dong H, Xu D.Z, Huang J.F and Sun C.P 2013 Coherent excitation transfer via the dark-state channel in a bionic system, Light: *Science & Applications.* 1 e2.

[14]Sen A. Lin C. J and Kaun C.C 2013 Single-Molecule Conductance through Chiral Gold Nanotubes, *J. Phys.* Chem. C 117 13676.

[15]Yang H.H, Chu Y.H, Lu C.I, Yang T.H, Yang K.J, Kaun C.C, Hoffmann X and Lin M .T 2013 Digitized Charge Transfer Magnitude Determined by Metal-Organic Coordination Number, *ACS Nano.* 7 2814.

[16]Croitoru M.D, Shanenko A. A, Kaun C.C and Peeters F.M 2012 Ultra-small metallic grains: effect of statistical fluctuations of the chemical potential on superconducting correlations and vice versa, *J. Phys.: Condens. Matter.* 24 275701.

[17]Sen A and Kaun C.C 2010 Effect of Electrode Orientations on Charge Transport in Alkanedithiol Single-Molecule Junctions, *ACS Nano.* 4 6404.

[18]Tang Y.H, Bagci V.M.K and Chen J.H 2011 Conductance of Stretching Oligothiophene Single-Molecule Junctions: A First-Principles Study, *J. Phys*. Chem. C 115 25105.
[19]Luzhbin D.A and Kuan C.C 2011 Bias-driven super-insulating state in prismane single-Molecule contacts, *Physica Status Solidi-Rapid Research Letters*. 5 265.
[20]Tang H, Dou K.P, Kaun C.C, Kaung Q and Yang S.H 2014MoSe2 Nanosheets and Their Graphene Hybrids: Synthesis, Characterizations and Hydrogen Evolution Reaction Studies, *J. Mater. Chem*. A 2 360.
[21]Hung Y.L 2017 Evaluation of the Circular Economy and Synergy Development Level of Cement Industry in Taiwan,National Cheng Kung University Department of Resources Engineering Thesis.
[22]Melinda S.L 2011 The Necessity to Exploit the Economic Network's Synergistic Potential, *LAP LAMBERT Academic Publishing*, Saarbrücken 33-37.
[23]World Intellectual Property Organization.
[24]Web Of Science.
[25]OECD Statistical database.
[26]International renewable energy agency.

IOP Conf. Series: Materials Science and Engineering **720** (2020) 012004 doi:10.1088/1757-899X/720/1/012004

Mechanical properties of the wet mixed geopolymer concrete based on CDG and Slag cementitious materials in ambient curing temperature

Xiaoxiong Zha*, Kai Wang, and Jiayuan Zhu

School of Civil and Environmental Engineering, Harbin Institute of Technology Shenzhen, Shenzhen, Guangdong, China.

zhahero@126.com, cewangkai@126.com, zhujiayuan1994@sina.com

Abstract. Although considerable research has been conducted on alkali-activated fly ash cementitious materials for geopolymer concrete production, few of which concentrated on the slag-based geopolymer concrete. This paper presents the results of an experimental investigation on the 7-day & 28-day compressive strength, as well as 28-day flexural strength of the wet mixed geopolymer concrete based on CDG and slag cementitious materials cured at ambient temperature. A total of 90 specimens were tested. The cementitious materials content, sand ratio, as well as the liquid-solid ratio were selected as the main parameters. Besides, the new alkali activator system was adopted to prepare the geopolymer. Effects of these variables on the compressive and flexural behavior of the wet mixed geopolymer concrete are plotted throughout this paper.

Keywords. Geopolymer concrete, Slag, Cementitious material, Compressive strength, Flexural Strength

1. Introduction

Recently, geopolymer concrete has shown great potential in the construction industry due to its environmentally friendly characteristics. The utilization of geopolymer concrete not only reduces the emission of greenhouse gases associated with the global warming problem (Davidovits,1994; Heath et al., 2014), but also exhibits excellent mechanical and durability performance over the OPC concrete in unregulated disposal of the industrial waste by recycling these materials in geopolymer production (Bernal et al., 2012; Ferone et al., 2015; Zhang, 2013). Since the conventional fly ash-based geopolymer concrete must be cured at high temperatures in the obtaining of sufficient early strength properties, the development of geopolymer concrete in engineering applications has been extremely limited (Mohammed et al., 2016). In this context, researches all over the world have been carried out on the calcium enriched slag cementitious materials. It is shown that the added calcium components could be helpful in solving the problem of geopolymer concrete being cured at ambient temperature (Suwan and Fan, 2014, 2016; Peng et al., 2019; Li et al., 2017; Yip et al., 2008). Since the completely decomposed granite (CDG) mainly consists of aluminosilicates, in which the alumina tetrahedron could react with the silica tetrahedron, under the activation effect of the alkali metal ion. The component CDG can be used as the alumina-silicate precursor in geopolymer manufacturing process.

Besides, study has already shown that the CDG can be used as the main cementitious material in geopolymer (Zha and Zhu, 2018).

In this study, mechanical properties of wet mixed geopolymer concrete cured at ambient temperature on considering the variation of the cementitious materials content, the sand ratio, and the liquid-solid ratio were conducted. Since the strong alkali sodium hydroxide could release large amounts of heat in the geopolymer manufacturing process, the calcium hydroxide and anhydrous sodium carbonate are adopted as the alkaline activator. Besides, the wet mixing technique was adopted to premix the water glass liquid, calcium hydroxide, anhydrous sodium carbonate, and water. The technique made it possible for the alkaline activator to be mixed evenly, which is helpful for the geopolymerization reaction.

2. Materials and methods

2.1. Cementitious materials

In the experiment, the completely decomposed granite (CDG) and the blast furnace slag were used as the main cementitious materials. Details of the cementitious materials are exhibited as follows.

2.1.1. Completely Decomposed Granite （CDG）

The completely decomposed granite (CDG) used in the test came from a construction site in the University Town of Shenzhen, Nanshan District, Shenzhen, Guangdong Province, China. This kind of soil widely exists in the Southern China. According to the Code for Highway Engineering Geological Investigation (JTG-C20-2011), the soil is defined as the quaternary loose deposits remaining in situ after the granite is weathered, called the completely decomposed granite residual soil. Since the soil is usually treated as the waste soil in the foundation excavation or in the site formation works, it belongs to one kind of waste soil. The CDG used in this test contains sand and gravel clay, showing a middling brown red color. After dried the excess liquid, the clod was carefully grinded into a smaller particle size and a higher fitness. This treatment made it easier for the chemical reaction between the CDG and the alkali-activator. The CDG sample used in the test is shown in Figure 1(a).

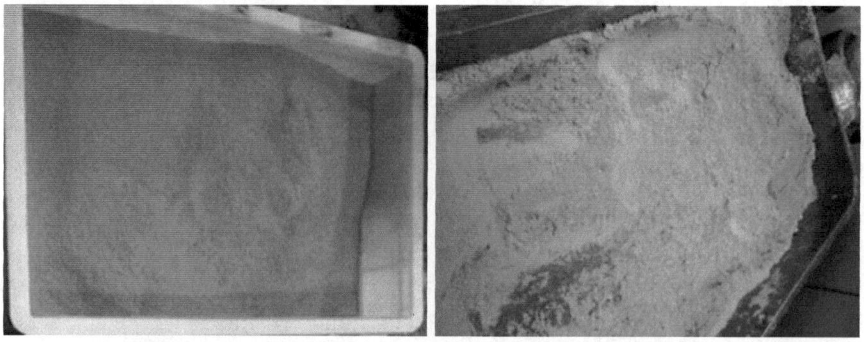

(a) Completely decomposed granite (b) Granulated blast furnace slag power

Figure 1. Cementitious materials used in the experiment

2.1.2. Blast Furnace Slag

The blast furnace slag used in the experiment was the S-95 grade granulated blast furnace slag powder, a by-product in steel-making industry, which was from Zhengzhou, Henan City. The slag powder sample is shown in Figure 1(b). Based on the X-ray fluorescence (XRF) analyses method, chemical composition of the cementitious materials (CDG and Slag) were successively obtained, and the main chemical components are exhibited in Table 1. It can be seen from the table that the CDG mainly contains SiO_2 and Al_2O_3, while the granulated blast furnace slag mainly contains SiO_2, Al_2O_3 and CaO.

Table 1. Chemical composition of the cementitious materials used in the experiment

Composition	SiO_2 (%)	Al_2O_3 (%)	Fe_2O_3 (%)	CaO (%)	MgO (%)	TiO_2 (%)	Na_2O (%)	K_2O (%)
CDG	45.71	36.92	8.96	0.49	-	4.40	-	2.96
Slag	31.57	15.27	0.23	43.18	6.68	0.74	0.45	0.21

2.2. Alkaline activator and Aggregates

The water glass solution used in the test was the liquid-4 type liquid water glass with a modulus of 2.31, in which the SiO_2 content is 29.84%, and the Na_2O content is 13.36%. The calcium hydroxide and the anhydrous sodium carbonate, which are in white color powder and water-soluble, were mainly used to adjust the modulus of the water glass solution. Alkaline activator samples were shown in Figure 2.

(a) Liquid Na_2SiO_3 (b) $Ca(OH)_2$ (c) Na_2CO_3

Figure 2. Alkaline activator used in the experiment

The coarse aggregate and fine aggregate used in the concrete are the crude gravel and river sand, from the Tanking Mountain in Shenzhen. The particle sizes of the crude gravel were from 5 mm to 25 mm, while the particle sizes of the river sand were from 0 mm to 5 mm.

3. Experimental methods

To solve the great heat released problem in the preparation of CDG geopolymer concrete, the alkali sodium hydroxide was replaced by the calcium hydroxide and anhydrous sodium carbonate, which was also used to verify whether the method was feasible. Besides, the wet mixing technique was adopted to premix the water glass liquid, calcium hydroxide, anhydrous sodium carbonate, and water. After the alkaline activator and water were mixed and stirred for 60 s in a uniform state, the alkaline activator solution and the concrete aggregate were poured together into the mixer. Subsequently, the mixing process was maintained for 120 s. After that, the stirred CDG geopolymer concrete was poured into a mold of 100 mm ×100 mm×100 mm with the help of an electric concrete vibrator. Then the fabricated specimen were cured in the natural environment maintenance condition for 7 days or 28 days.

Determination of the specimen's flexural strength (R_f) and the specimen's compressive strength (R_c) are illustrated in Figure 3(a) and 3(b). The calculation methods are exhibited as follows.

$$R_f = \frac{2 \times F}{\pi \times A} = 0.637\frac{F}{A} \quad (1)$$

$$R_c = \frac{F_c}{A} \quad (2)$$

In which, F is the ultimate load when the specimen is broken, A is the cross-section area of the specimen, R_f is the specimen's flexural strength, R_c is the specimen's compressive strength.

(a) Compressive strength (b) Flexural strength

Figure 3. Strength measurement methods

3.1. Mix ratio design of the cementitious materials content

Three groups containing 27 specimens were cast, and were labelled as C1-1, C1-2, and C1-3. The sand ratio and the liquid-solid ratio remain constant, which equal to 40% and 0.65 respectively, while the cementitious materials contents are 420 kg/m^3, 450 kg/m^3, and 480 kg/m^3. Design of the mixture ratio was illustrated in Table 2 as follows.

Table 2. Mix ratio design of the cementitious materials content

Label	CDG (g)	Slag (g)	$Ca(OH)_2$ (g)	Na_2CO_3 (g)	Na_2SiO_3 (g)	Fine aggregate (g)	Coarse aggregate(g)	H_2O (g)
C1-1	2520	1680	112	159	1299	6800	10200	1160
C1-2	2700	1800	120	170	1390	6800	10200	1245
C1-3	2880	1920	128	181	1481	6800	10200	1330

3.2. Mix ratio design of the sand ratio

Accordingly, the 27 specimens in this part were classified into three groups, which were labelled as C2-1, C2-2, C2-3. The cementitious materials content and the liquid-solid ratio remain constant, which equal to 450 kg/m^3 and 0.7. The sand ratio varies from 34% to 40%. Details of the mix ratio design were exhibited in Table 3 as follows.

Table 3. Mix ratio design of the sand ratio

Label	CDG (g)	Slag (g)	$Ca(OH)_2$ (g)	Na_2CO_3 (g)	Na_2SiO_3 (g)	Fine aggregate (g)	Coarse aggregate(g)	H_2O (g)
C2-1	2700	1800	130	185	1500	5780	11220	1335
C2-2	2700	1800	130	185	1500	6290	10710	1335
C3-3	2700	1800	130	185	1500	6800	10200	1335

3.3. Mix ratio design of the liquid-solid ratio

In this section, four groups containing 36 specimens were cast and labelled as C3-1, C3-2, C3-3, C3-4. The cementitious materials content and the sand ratio remain constant, which equal to 450 kg/m^3 and 40%, respectively. The liquid-solid ratio varies from 0.65 to 0.80. Details of the mix ratio design were shown in Table 4 as follows.

Table 4. Mix ratio design of the liquid-solid ratio

Label	CDG (g)	Slag (g)	$Ca(OH)_2$ (g)	Na_2CO_3 (g)	Na_2SiO_3 (g)	Fine aggregate (g)	Coarse aggregate(g)	H_2O (g)
C3-1	2700	1800	120	170	1390	6800	10200	1245
C3-2	2700	1800	130	185	1500	6800	10200	1335
C3-3	2700	1800	140	200	1610	6800	10200	1425
C3-4	2700	1800	150	215	1720	6800	10200	1515

4. Results and discussions

4.1. Influence of the cementitious materials content

After the natural curing was carried out, the specimens were subsequently subjected to the compression tests or flexural tests. Figure 4 clearly presents the 7-day, 28-day compressive strength and the 28-day flexural strength of specimens, considering the variation of the cementitious materials content.

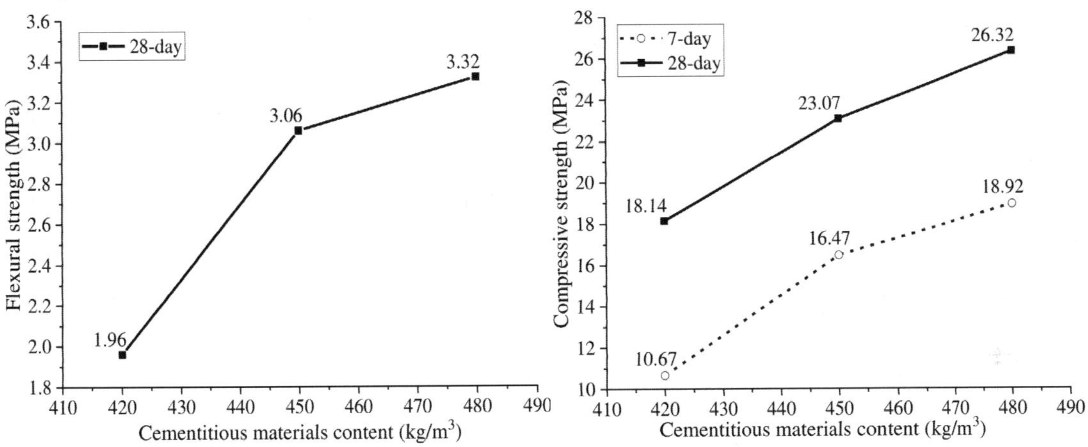

(a) Flexural strength (b) Compressive strength

Figure 4. Influence of the cementitious materials content on the specimen's strength

It can be seen from Figure 4(a) that as the cementitious materials content increases from 420 kg/m^3 to 450 kg/m^3, the flexural strength increases significantly. When it increases from 450 kg/m^3 to 480 kg/m^3, the flexural strength increases slightly, and the maximum 28-day flexural strength is achieved when the cementitious materials content is 480 kg/m^3. These can be explained by the fact that within certain limits the higher cementitious materials contain more SiO_2 and Al_2O_3, the chemical compositions which are helpful for forming more silica tetrahedron and aluminium-oxygen tetrahedron. Therefore, the more stable and complex three-dimensional spatial network structures are easier to be formed. Figure 4(b) shows that, compared with the flexural strength, the compressive strength increases a little slower as the cementitious materials content develops, regardless of the 7-day compressive strength or the 28-day compressive strength. In addition, the compressive strength increases with the curing time, and the 7-day compressive strength was about 70% of the 28-day compressive strength.

4.2. Influence of the sand ratio

The sand ratio has effects on the concrete aggregate gradations and influences the workability and strength of concrete. Regardless of the content of the sand ratio, the fluidity would become poor and the segregation would be occurred. Consequently, the mechanical properties of the geopolymer concrete were deteriorated. Figure 5 clearly shows the 7-day, 28-day compressive strength and the 28-day flexural strength of specimens, when the variation of the sand ratio was taken into account.

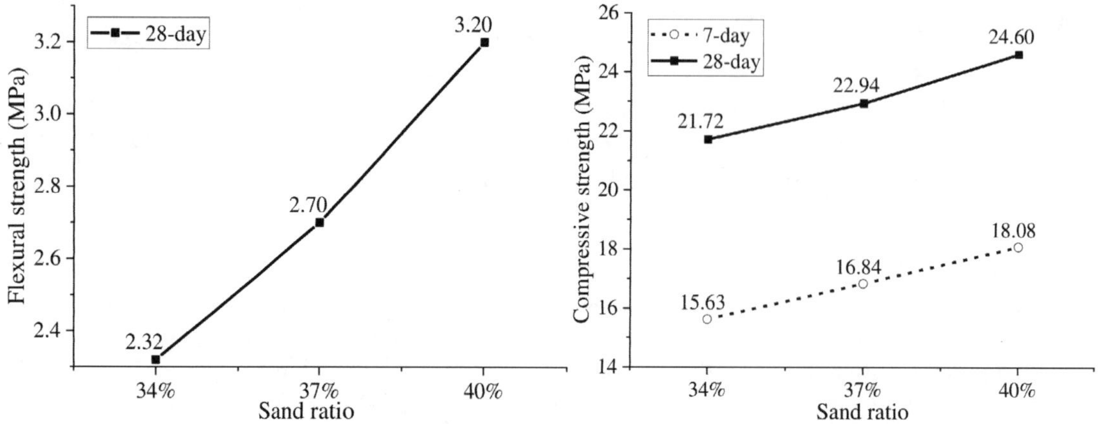

(a) Flexural strength (b) Compressive strength

Figure 5. Influence of the sand ratio on the compressive strength and flexural strength

Figure 5(a) indicates that the 28-day flexural strength increases remarkably with the sand ratio ranging from 34% to 40%, and the relationship is approximately linear. Figure 5(b) shows that the compressive strength increases slightly as the sand ratio increases. Additionally, the ultimate 7-day and 28-day compressive strength are achieved when the sand ratio is 40%, which equal to 18.08MPa and 24.60MPa, respectively. The sand ratio has little effects on the compressive strength. When it increases from 34% to 40%, the compressive strength only improves 11.70%. The sand ratio influences the workability of the geopolymer concrete, this can be explained by the fact the sand could fill the gap of the concrete and the free water content would decreases. Consequently, the slump of the geopolymer is improved.

4.3. Influence of the liquid-solid ratio

The cementitious materials content and the sand ratio per cubic meter remain constant, while the liquid-solid ratio changes from 0.65 to 0.80. Figure 7 displays the 7-day, 28-day compressive strength and the 28-day flexural strength of specimens, when the variation of the liquid-solid ratio was considered.

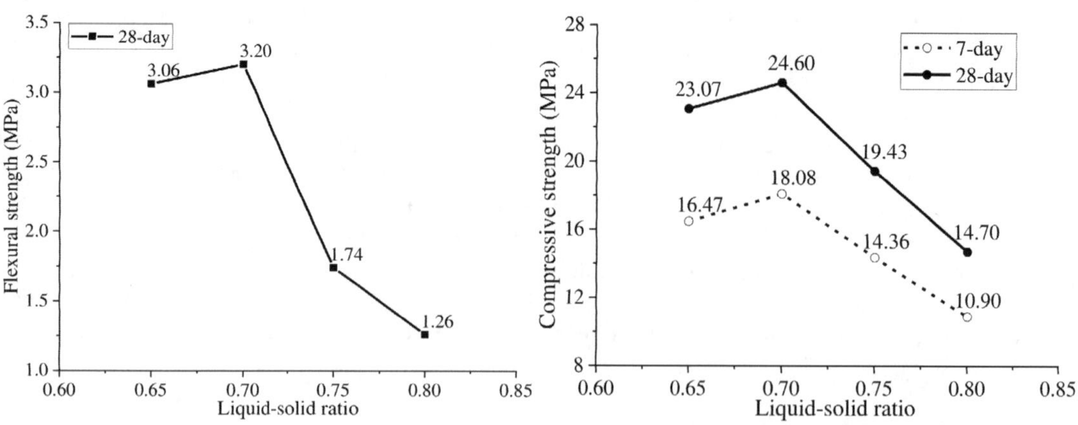

(a) Flexural strength (b) Compressive strength

Figure 6. Influence of the liquid-solid ratio on the compressive strength and flexural strength

Figure 6(a) shows that the flexural strength and the compressive strength increase firstly and then decrease, as the liquid-solid ratio increases from 0.65 to 0.80, regardless of the natural curing duration. When the liquid-solid ratio is 0.70, the maximum concrete strength is achieved. The 28-day flexural strength is 3.20MPa, the 7-day and 28-day compressive strength are 18.08MPa and 24.60MPa, respectively. Additionally, the 7-day compressive strength is approximately 70% of the 28-day compressive strength. The liquid-solid ratio directly affects the geopolymer concrete strength. With the increases of the liquid-solid ratio, the hydration product silica gel could fill the porous space and the density of the concrete is further dense. When a high proportion of the liquid-solid ratio exists, evaporation of the water from the alkaline activator solution results in pore structures, thus leading to a decline in the geopolymer concrete strength. When the liquid-solid ratio is relatively low, the alkaline activator solution could not react well with the cementitious materials. The reaction of the condensation polymerization is insufficient, which results in the reduction of the geopolymer concrete strength.

5. Conclusion

In this study, mechanical properties of wet mixed geopolymer concrete on considering the variation of the cementitious materials content, the sand ratio, and the liquid-solid ratio were conducted. It can be concluded that the flexural strength and the compressive strength both increase as the cementitious materials content increases from 420 kg/m^3 to 480 kg/m^3, when the sand ratio and the liquid-solid ratio remain constant. The flexural strength and the compressive strength both increase as the sand ratio increases from 34% to 40%, with the cementitious materials content and the liquid-solid ratio being constant. Moreover, the flexural strength and the compressive strength increase firstly and then decrease, as the liquid-solid ratio increases from 0.65 to 0.80, with the cementitious materials content and the sand ratio being constant. When the sand ratio is 0.70, the 7-day maximum compressive strength and the 28-day maximum compressive strength are achieved which equal to 18.08MPa and 24.60MPa respectively, while the 28-day maximum flexural strength is 3.20MPa.

Acknowledgement

This research is funded by the National Nature Science Foundation of China (Grant No. 51578181), Natural Science Foundation of Guangdong Province (No. 2016A030313665) and Shenzhen Science and Technology Plan Project (No. JCYJ20150327155221857). Acknowledgement is also given to the Shenzhen Carbon Storage Cement-Based Materials Engineering Laboratory.

References

[1] Davidovits J. (1994). "Global warming impact on the cement and aggregates industries," World Resource Review, 6:2.
[2] Heath A., Paine K., and McManus M. (2014). "Minimising the global warming potential of clay based geopolymers," Journal of Cleaner Production, vol. 78, pp. 75-83.
[3] Bernal SA., de Gutiérrez, RM., and Provis JL. (2012). "Engineering and durability properties of concretes based on alkali-activated granulated blast furnace slag/metakaolin blends," Construction and Building Materials, vol. 33, pp. 99-108.
[4] Ferone C., Liguori B., Capasso I., Colangelo F., Cioffi R., Cappelletto E.,and Di Maggio R. (2015). "Thermally treated clay sediments as geopolymer source material," Applied Clay Science, vol. 107, pp. 195-204.
[5] Zhang L. (2013). "Production of bricks from waste materials–a review," Construction and Building Materials, vol. 47, pp. 643-55.
[6] Mohammed Haloob Al-Majidi, Andreas Lampropoulos, Andrew Cundy, Steve Meikle. (2016) "Development of geopolymer mortar under ambient temperature for in situ applications," Construction and Building Materials, Volume 120, Pages 198-211.
[7] Suwan T and Fan M. (2014). "Influence of OPC Replacement and Manufacturing Procedures on the Properties of Self-Cured Geopolymer," Construction & Building Materials, 73(2):551-

561.
[8] Suwan T, Fan M, Braimah N. (2016). "Internal Heat Liberation and Strength Development of Self-Cured Geopolymers in Ambient Curing Conditions," Construction & Building Materials, 114(7):297-306.
[9] Peng H, Cui C, Liu Z, et al. (2019). "Synthesis and Reaction Mechanism of an Alkali-Activated Metakaolin-Slag Composite System at Room Temperature," Journal of Materials in Civil Engineering, 31(1).
[10] Li N , Shi C , Wang Q , et al. (2017). "Composition design and performance of alkali-activated cements," Materials and Structures, 50(3).
[11] Yip C K, Lukey G C, Provis J L, et al. (2008). "Effect of calcium silicate sources on geopolymerisation," Cement and Concrete Research, 38(4):554-564.
[12] Zha X, Zhu J, "Experimental Study on the Effects of Calcium-rich Materials on the Early Compressive Strength of Completely Decomposed Granite (CDG) Base-geopolymer"[C]//4th International Symposium on Hydrogen Energy, Renewable Energy and Materials, HEREM 2018, 2018:2-8.
[13] JTG-C20-2011, Code for Highway Engineering Geological Investigation, China.

Improve subsequent leaching efficiency and extraction rate of germanium in optical fibre cables with pre-treatment

Wei-Sheng Chen, Bi-Cheng Chang, Chih-kang Shuai
Department of Resources Engineering, National Cheng Kung University, No.1, University Road, Tainan City, Taiwan, Republic of China

Chihknag861117@gmail.com

Abstract. Germanium is an important and rare material which widely used in fibre optic industry, accounting for about 40% of the total germanium product usage. Due to the rarity of germanium, developing an effective recycling process of recovering germanium metal from waste fibre optical cables is significant. Fibre optical cables was arranged to do characteristic analysis, pre-treatment and dissolution in this study. In the characteristic analysis, the content of germanium dioxide in the glass fibre was 0.15%, and the remaining components were all silicon dioxide. In the pre-treatment part, it consists of two steps which are the solvent treatment and roasting. At the solvent treatment step, glass fibre can be separated by acetone with a concentration of more than 50% and ethanol with concentration of more than 75%. At the roasting step, NaOH was used as a roasting additive. Adding 5 molar ratio of NaOH/SiO2 at 500° C for 2 hours can totally convert the optical fibre into the silicate. After roasting, it could be leaching by dilute H2SO4 and its leaching rate is higher than 99.5%. This study can provide a process to recovery germanium from optical fibre effectively. According to the optimal conditions, the recovery of germanium could be up to 99%.

Key word：Germanium, Waste fibre optical, Solvent treatment, Roasting.

1. introduction

Germanium is a rare metal that occurring to the range of 1-7 ppm [1] and it also is a useful element in optics and electronics. According to the U.S. Geological Survey research, the world reserve of germanium only has 8600 tons [2] and it has none in mineable deposits. With the increased of demand for the semiconductor industry and the optical industry, the European Union has listed germanium as a critical raw material in 2010 [3]. And it is also an indispensable strategic resource in the modern military field. According to statistics, about 30% of the world's use of germanium comes from recycled materials, almost electronic devices. In the global consumer market, the part used for optical fiber accounts for about 40%. Infrared optics accounted for 30% [4]. With the development of 5G networks and advances in optical fiber technology, as the main added material of optical fiber, the demand for it will inevitably continue to rise, and lots of discarded optical fiber cables will also be generated. Germanium as the scarce metals, the future resources will face a shortage of supply. Therefore, the disposal of these discarded optical fiber cables is necessary, and because of the limited resources of germanium, it is necessary to study the recovery process of germanium in waste fibers.

At present, there is only a recycling process for germanium in optical fiber process wastewater [5], and there is no treatment process for the complete set of waste optical fibers. Waste optical fiber cables require pre-treatment, including stripping, crushing, solvent treatment, roasting, pyrolysis, and more. In this study, the solvent was use and the waste fiber was treated by roasting. There are two main methods for recovering germanium from treated optical fiber cables, and pyrometallurgy and hydrometallurgy. The pyrometallurgical process is most commonly used to recover germanium from coal [5-9], but it has high impurity content and energy consumption. Larger, hydrometallurgy on treated fibers is more suitable than pyrometallurgy. Common hydrometallurgical processes for the recovery of germanium include leaching chemical precipitation [10-11], chlorination distillation [12], flotation [13], ion exchange [14-15], solvent extraction [16]. In order to effectively improve the efficiency of subsequent germanium separation, this study chose the impregnation method to recover the germanium in the waste optical fiber cable.

optical fiber cables are stripped by stripping machine, and then, the optical fiber cables were processed through two pre-treatments, solvent treatment and roasting. In order to improve the subsequent leaching rate, in the solvent treatment part, the separation effect of different solvents, concentrations and time on the plastic was studied. In the roasting part, different fluxes, temperatures, and dosing amounts were investigated. The results show that two different types of glass fibers and sodium silicate type, which are in the form of silicon dioxide, are formed. In this study, the sodium silicate was selected as the subsequent acid leaching analysis. According to the research, better leaching rate can be obtained under the optimal parameters.

2. Experimental

2.1. Materials, reagents and instruments

In this study, the optical fibre samples were collected from waste cables. According to the previous study, the compounds analysis of optical fibre cable was shown in table 1, the elements analysis was shown in Table 2. which could be found the concentration of germanium in the fibre was 0.1% and other is silica. Figure 1. Show that the XRD and SEM/EDS analysis of optical fibre powder. The acetone (ECHO C_3H_6O 99.5%) and ethanol (ECHO C_2H_5OH 99.5%) were used to separate the plastic on the optical fibre cable. The sodium hydroxide (Applichem Panreac NaOH 98%) and sulfuric acid (Sigma-Aldrich H_2SO_4 98%) used as roasting and leaching agents and dissolved in deionized water. Analysing the metal content for separation and leach efficiency is used by Inductively Coupled Plasma Optical Emission Spectrometry (ICP-OES, Varian, Vista-MPX). Using Scanning Electron Microscope (SEM, Hitachi, S-3000N), X-ray Fluorescence (XRF, SPECTRO XEPOS), X-ray diffraction (XRD, Dandong, DX-2700) and Energy Dispersive X-ray Spectroscopy (EDS) to get information about the sample's surface topography and composition.

Table 1. Compounds analysis of optical fibre cable

Aluminum strip	**Plastic Jacket**	**Optical Fiber**	**Steel wire**	**Filling compound**	**Fiber bundle**
5~8%	20~25%	5~10%	15~20%	10~15%	2~5%

Table 2. Elements analysis of optical fibre powder

Elements	Ge	Silica	Fe	Mg	Ca
XRF analysis	0.11%	99.48%	0.23%	0.04%	0.13%
ICP analysis	0.1%	99.9	Not-detected	Not-detected	Not-detected

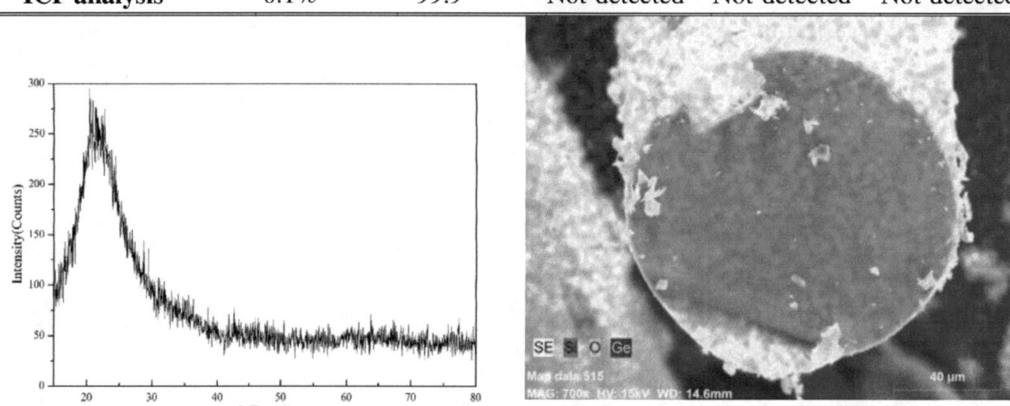

Figure 1. XRD and SEM/EDS analysis of optical fiber powder

IOP Conf. Series: Materials Science and Engineering **720** (2020) 012005 doi:10.1088/1757-899X/720/1/012005

2.2. Plastic softening

The purpose of this part is to process the fiber bundle tube containing germanium. At first, optical fiber cables are stripped by stripping machine. After the discarded fiber optic cable is stripped, the external plastic is removed, and the remaining product is a fiber bundle tube that cannot be physically separated. This is the goal of this study. The fiber bundle tube is composed of glass fiber and acrylic resin coating. Therefore, the use of plastics can be softened by organic solvents, solvent treatment is beneficial to the subsequent hydrometallurgical process. The organic solvent part is selected from 10%, 25%, 50%, 75%, 90% acetone and ethanol as a separating agent to investigate the separation effect of the concentration and time on the plastic.

2.3. alkali roasting

First, the optical fibre was collected from cables. Because the fibre contained silicon dioxide [17], it should dissolve silica and take germanium out of them by alkali roasting. In the alkali roasting part, alkaline fluxes such as sodium carbonate, potassium hydroxide and sodium hydroxide are selected, and the optimum calcination temperature, the selection and addition amount of the agent are discussed. According to the literature, the roasting temperature is selected as 300° C, 400° C, 500° C, 600° C, 700° C, 800° C, 900° C as parameters. The dosage is the ratio of the agent to the silicon dioxide molar, and 2, 3, 4, 5, 6, 7, 8 mol / mol is used. The roasting time was two hours and the roasting efficiency was impregnated with deionized water and compared with the aqua regia and HF total dissolution samples, and the crystal phase change was analysed by XRD to provide a set of optimum roasting conditions.

2.4. Leaching

The leaching efficiency of two types of silicon dioxide is discussed. For the solvent-treated fibre, HNO_3, HCl and H_2SO_4 were used as the main leaching liquid, and the leaching efficiency under the addition of HF was investigated. Then discuss the effect of liquid-solid ratio and time on the leaching efficiency. For the alkali roasted fibre, and deionized water is used as a leaching liquid, and H_2SO_4 is used as an acid-base adjusting agent. The influence of pH and liquid-solid ratio on the leaching effect is sequentially discussed. The pH value was set from 1 to 7 and the effect of liquid-solid ratio from 10 to 100 ml/g which were tested to get better leaching efficiency [17].

3. Results and discussion

3.1. Plastic softening

The organic solvent can be used to soften the plastic, and 10%, 25%, 50, 75%, 90% acetone and ethanol are used as the separation agent, and the respective separation time is recorded in Table 2. According to the results, the acetone separation effect is much larger than that of ethanol, and the acetone has no peeling effect at a concentration of 50% or less, and the ethanol has a effect of less than 75%. After the separation is completed, the quality of the plastic and glass fibre in the bundle of fibre bundles is analysed. The results are listed in Table 3. According to the results, the plastic part accounts for about 57% of the bundle weight and the glass fibre accounts for about 43%.

Table 3. Separation effect of organic solvent on fiber bundle cluster coated plastic

acetone	10%	25%	50%	75%	90%
time	x	x	5 hours	2 hours	20minutes
ethanol	10%	25%	50%	75%	90%
time	x	x	x	6 hours	2 hours

Table 4. Weight analysis of fiber bundles

	Coated plastic	Glass fiber
Weight %	57.14%	42.86%

3.2. alkali roasting

According to the composition analysis of the glass fibre, silicon dioxide is the main component of the glass fibre. In this step, the roasting of the silicon dioxide at different temperatures of the alkaline flux of Na_2CO_3, KOH and NaOH is investigated. As a result, the roasting time was 2 hours, and the results are shown in Table 4. The results show that at 900 ° C, the addition of Na_2CO_3 can completely react and completely dissolve, KOH is 400 ° C and NaOH is 500 ° C, which is consistent with the literature reaction temperature [18-19].According to the above conclusions, the dosage of the flux at the optimum operating temperature is compared with the roasting efficiency. The dosing amount is the ratio of the agent to the silicon dioxide molar. The roasting efficiency is impregnated with deionized water and compared with the aqua regia. The results are shown in Figure 2. According to the results, it can be found that KOH and NaOH can reach complete reaction when the molar ratio is 5, and Na_2CO_3 can be completely reacted when the molar ratio is 8. The temperature and the required amount of the drug can be compared to find that the Na_2CO_3 efficiency is lower than KOH. With NaOH, and because of the amount of flux (the molecular weight of KOH is higher than NaOH) and the price of the flux (effective for NaOH), NaOH is selected as the best alkali roasting additive here. The optimal operating conditions are temperature 500 ° C, dosing amount. For 5 molar ratio, the roasting time was 2 hours. Finally, the product obtained under the optimum conditions was analysed by XRD, and the results are shown in Figure 3. According to the map analysis, it can be confirmed that under this roasting condition, the silicon dioxide has been converted into water-soluble sodium silicate, and subsequent leaching can be performed

Table 5. Alkali roasting effect of different agents at different temperatures

temperatures	300°C	400°C	500°C	600°C	700°C	800°C	900°C
Na2CO3	×	×	×	×	×	×	○
KOH	×	○	○	○	○	○	○
NaOH	×	×	○	○	○	○	○

(× : silicon dioxide does not fully react ; ○ : completely reacts and dissolves)
(roasting time: 2 hour)

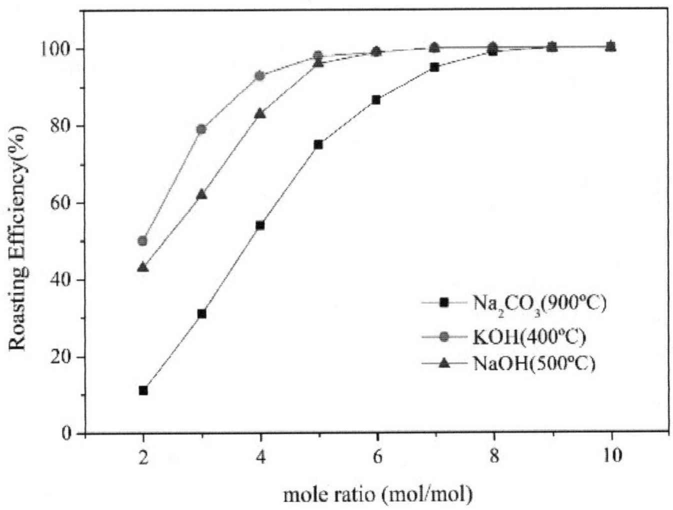

Figure 2. Effect of the roasting efficiency on mole ratio

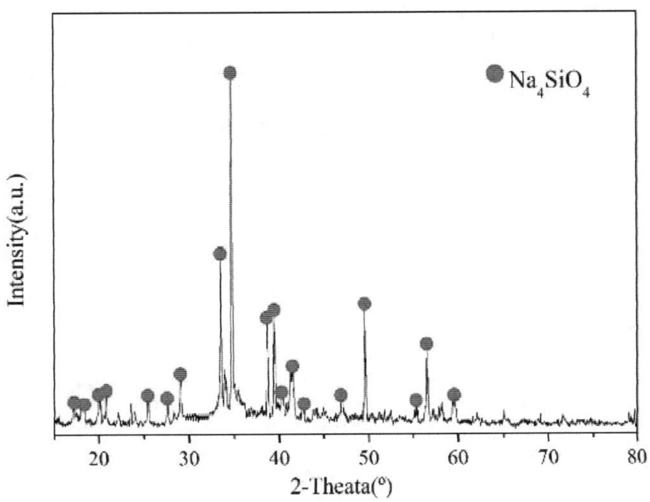

Figure 3. The XRD pattern of optical fibre after roasting.

3.3. Leaching

For the type of silicon dioxide, the effect of leaching efficiency on concentration and HF of different acids was investigated. The results are shown in figure 4. The operating conditions is L/S=100, 5%v/v HF, time=3hr. The results show that if HF is not added, the germanium coated in the silicon cannot be leached. For this reason, HF must be added. In terms of acid concentration, the impregnation rate of ruthenium will decrease with concentration at a certain acid concentration, and the results are in accordance with the literature. Therefore, the concentration is selected to be 0.1M H2SO4. According to the above, HF is necessary, so the influence of the concentration on the leaching efficiency is examined, and the results are shown in figure 5. The results show that the leaching efficiency of germanium rises as the HF concentration rises, so 5% v/v HF is selected. Using the optimized leaching liquid and concentration, the effect of leaching efficiency on liquid-solid ratio was investigated. The results are shown in figure 6. The results show that when the liquid-solid ratio is

IOP Conf. Series: Materials Science and Engineering **720** (2020) 012005 doi:10.1088/1757-899X/720/1/012005

reduced, in other words, when the leaching liquid is reduced, the driving force of the reaction is lowered, resulting in leaching efficiency of germanium drops. So L/S=100ml/g is selected. Finally, the effect of leaching efficiency on leaching time was investigated by using the above parameters. The results are shown in figure 7. The results show that the reaction reaches equilibrium at 3hr.

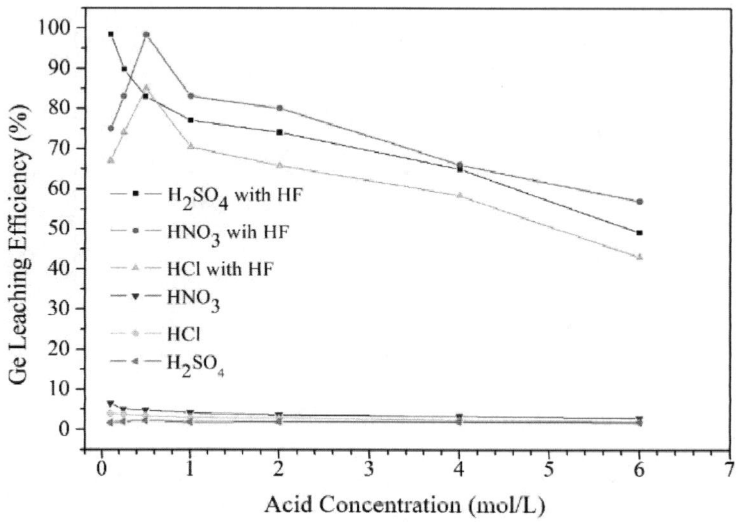

Figure 4. Effect of the leaching efficiency on concentration and addition of HF. (L/S=100, 5%v/v HF, time=3hr)

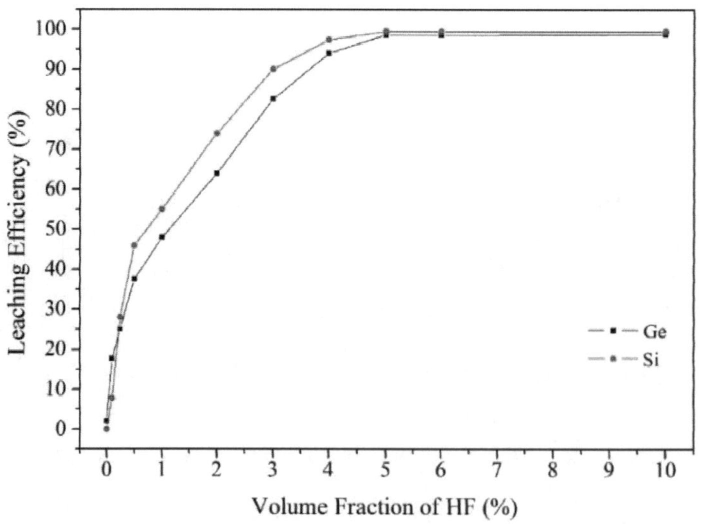

Figure 5. Effect of the leaching efficiency on concentration of HF. (L/S=100, 0.1M H2SO4, time=3hr)

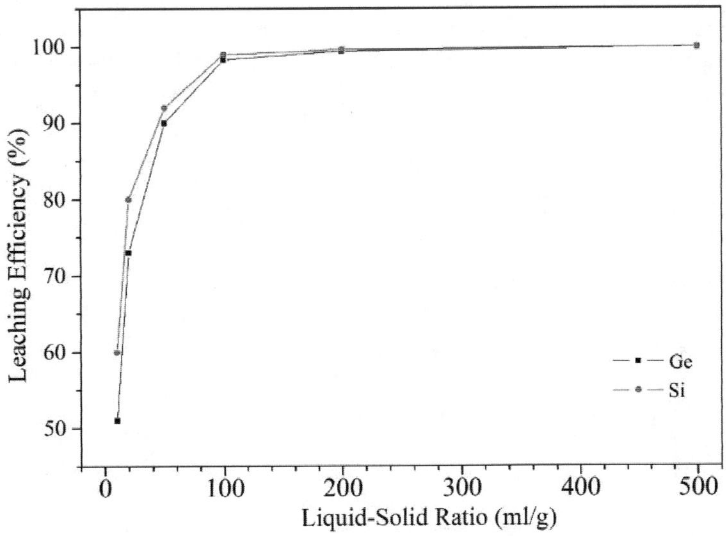

Figure 6. Effect of the leaching efficiency on liquid-solid ratio. (5%v/v HF, 0.1M H2SO4, time=3hr)

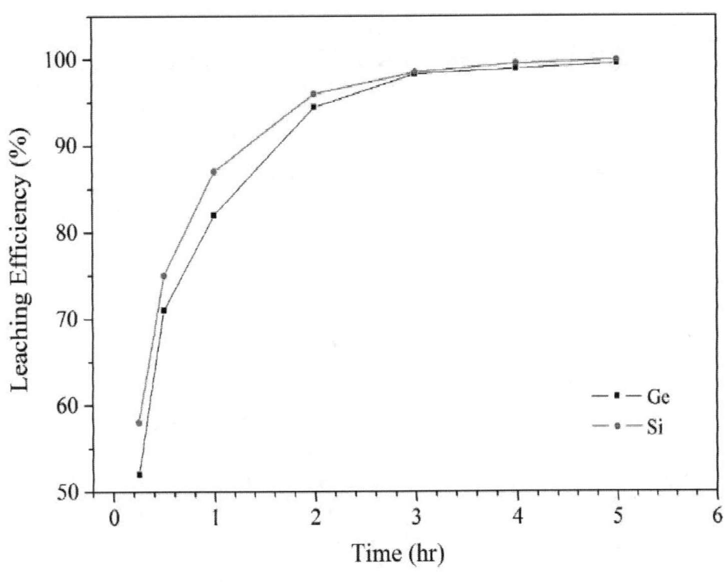

Figure 7. Effect of the leaching efficiency on leaching time. (L/S=100, 0.1M H2SO4, 5%v/v HF)

According to previous research [17], after roasting, the fibre was leached by dilute H_2SO_4 and showed the pH value and liquid-solid ratio effect in the figure 8 and 9. It was found that germanium and silicon leaching rate decreased significantly at neutral environment in figure 8. The reason of that if the image of the liquid and solid leaching rate were decreased with the liquid-solid ratio. The excessive silicon concentration would also make them precipitate that the leaching efficiency decreased.

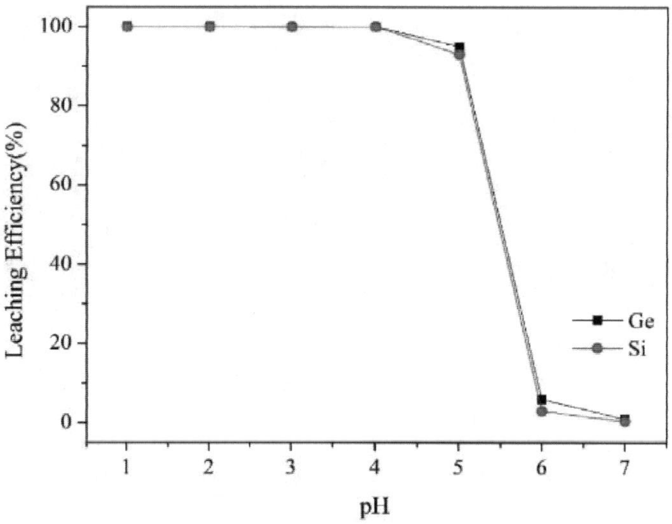

Figure 8. Effect of the leaching efficiency on pH value.
(pH=3, 25°C).

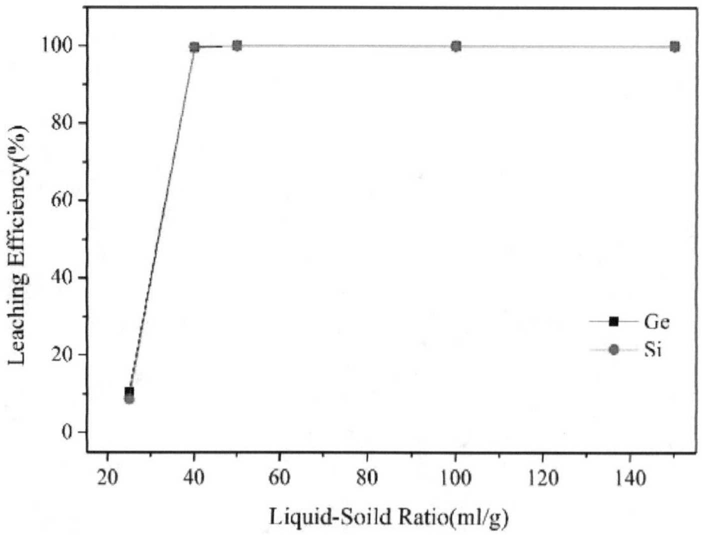

Figure 9. Effect of the leaching efficiency on liquid-solid ratio.
(L/S=40ml/g, 25 ◦ C).

4. Conclusion

This study proposes a hydrometallurgical method that can effectively improve the subsequent leaching efficiency and extraction success rate of germanium in optical fibre cables with pre-treatment. In the characteristic analysis part, the bundle of optical fibre is composed of an acrylic resin coating and a glass optical fibre. The plastic part accounts for 57.14% of the bundle of optical fibre and the glass fibre accounts for 42.86%. The analysis of the glass fibre can be found to be composed of amorphous silicon dioxide. Germanium is embedded in silicon dioxide, and the content of germanium dioxide is 0.15%, and the remaining components are silicon dioxide. In the part of separated the acrylic resin coating by the organic solvent, the plastic coating can be removed by using acetone having a concentration more than 50% or ethanol more than 75%, and the efficiency is preferably acetone. In the alkali roasting part, NaOH is selected as the optimum roasting additive. The additive can convert the glass fibre into sodium silicate at a temperature of 500 ° C, Adding 5 molar ratio of NaOH / SiO_2, and the roasting time of 2 hours. In the optimal leaching parameter part of sodium silicate type, adding 5% v/v HF, under the conditions of liquid-solid ratio 100 ml/g, reaction time 3h and 25°C , the efficiency can reach 99.5%. In the optimal leaching parameter part of sodium silicate type, deionized water is used as the substrate, and H_2SO_4 is used as the acid agent. The leaching is performed under the conditions of pH 1-4, liquid-solid ratio 40 ml/g, reaction time 20 minutes and 25 °C. The efficiency is greater than 99.5%. Comparing the two types of leaching, alkali roasting requires the addition of alkali agent and high temperature treatment, but less leaching liquid can be used in the leaching part, faster reaction time and avoiding the addition of HF. Therefore, this study suggests the choice of alkali roasting as the preferred method. This study shows a process of pre-treatment which can effective improve subsequent leaching efficiency and extraction success rate of germanium in optical fibre cables

References

[1] Adams, J. H., 1990, Germanium and Germanium Compounds. Metals Handbook, 10th ed.; ASM International: Materials Park.
[2] USGS, Kelly, T.D., and Matos, G.R., 2013, comps, Historical statistics for mineral and material commodities in the United States (2013 version): U.S. Geological Survey Data Series 140.
[3] European Commission, 2010, "Critical Raw Materials for the EU: Report of the Ad-Hoc Working Group on Defining Critical Raw Materials," European Commission: Brussels, Belgium.
[4] U.S. Geological Survey, 2017, "Mineral Commodity Summaries 2017."
[5] M. Bohrer et al., "A process for recovering germanium from effluents of optical fiber manufacturing," Journal of Lightwave Technology, 3(3), 699-705, 1985.
[6] Z. Rong-Kun, Extraction of germanium from lignite, CN1101380 A, 1995.
[7] L. Ting et al., Method for preparing carbocoal and extracting germanium-contained matter from brown coal destructive ditillation, CN1814701 A, 2006.
[8] L. Ba-Ke, G. Chi -Ming, Method for extracting germanium from lignite by pyrogenic process, CN101413063 A, 2009.
[9] P. Shih-Kun et al., Germanium ore washing method capable of improving pyrometallurgy recovery rate, CN105170311 A, 2015.
[10] F. Arroyo et al., "Recovery of Germanium from Coal Fly Ash Leachate by Precipitation," World of Coal Ash (WOCA) Conference, May, 2009.
[11] F. Arroyo et al., "Precipitation of Germanium from Coal Fly Ash Leachates," Coal Combustion and Gasification Products, 2(1), 28–34, 2010.
[12] J. Jandová et al., "Recovery of germanium from fly ash-waste from coal combustion," In Proceedings of European Metallurgical Conference, Fridrichshafen, BRD; GDMB: Clausthal, 69-75, 2002.
[13] Y. Inukai et al., "Selective separation of germanium(IV) by 2,3-dihydroxypropyl chitosan resin," Analytica Chimica Acta, 371(2-3), 187-193, 1998.
[14] F. J. S. Menendez et al., "Process for the recovery of germanium from solutions that contain it," U.S. Patent 4886648, 1989.
[15] D. A. D. Boateng, D. L. Ball, G. M. Swinkels, "Recovery of germanium from aqueous solutions," US Patent 4525332, 1989.
[16] G. Cote, D. Bauer, "Liquid-liquid extraction of germanium with oxine derivatives," Hydrometallurgy, 5(2-3), 149-160, 1980.
[17] W. S. Chen, B. C. Chang, K. L. Chiu, 2017, "Recovery of germanium from waste Optical Fibers by hydrometallurgical method," Journal of Environmental Chemical Engineering, 5(5), 5215–5221.
[18] H. MORI, "Extraction of Silicon Dioxide from Waste Colored Glasses by Alkali Fusion Using Sodium Hydroxide," Journal of the Ceramic Society of Japan, 111(1294), 376-381, 2003.
[19] H. Mori, "Extraction of silicon dioxide from waste colored glasses by alkali fusion using potassium hydroxide," Journal of Materials Science, 38(16), 3461-3468, 2003.

Study on factors affecting the mechanical properties of dry mixing CDG based geopolymer

Xiaoxiong Zha*, Kai Wang, and Jiayuan Zhu

School of Civil and Environmental Engineering, Harbin Institute of Technology Shenzhen, Shenzhen, Guangdong, China.

zhahero@126.com, cewangkai@126.com, zhujiayuan1994@sina.com

Abstract. Cement is the main material for the construction industry. But it produces a large amount of greenhouse gases and consumes a lot of resources. As a new type of green cementitious material, geopolymer materials can replace cement. In order to simplify the preparation process of the geopolymer material, dry-mixing CDG geopolymer materials are prepared by using a dry mixing process technique, using a solid activator material, mixing the dry materials, and adding water to the mixture. In this paper, the completely decomposed granite (CDG) and slag are used as the main cementitious materials, and the geopolymer paste and concrete are prepared by the action of alkaline activator, and the systematic study on the mechanical properties of this new geopolymer material has been carried out.

Keywords. Geopolymer, CDG, Slag, Dry mixing, Mechanical properties

1. Introduction

Cement is the main material of the construction industry, and emits a large amount of carbon dioxide. Geopolymer materials can be used as a new type of green cementitious material to replace some or all of the cement (Davidovits, 1994). Since the introduction of geopolymer by Davidovits, after decades of research and development, the selection of raw materials for production has gradually turned into solid industrial waste (Shi et al., 2013). This not only realizes the secondary utilization of industrial waste, but also greatly reduces the cost in the production process of geopolymer materials. As researchers continue to apply and develop this material, the prices of industrial waste are also rising. Some studies have shown that completely decomposed granite can be used as a raw material for geopolymer, so CDG is used as the main cementitious material in this research (Zha et al., 2018).

The preparation of geopolymer materials under the common process requires the preparation of an activator solution, which leads to a cumbersome preparation process; and the activator solution is corrosive, which also increases safety hazards during the configuration process. This also has some adverse effects on the promotion and development of geopolymer materials at the construction site. In response to this problem, Suwan developed a self-maintenance technology that uses a "pre-dry mix" method (Suwan et al., 2014, 2016) based on the exothermic heat of the geopolymer activator dissolved in water to provide the heat needed for strength development, allowing the geopolymer material to be at ambient temperature. Under the shape can be cured. Bayuaji also proposed a new idea for the production of geopolymer cement technology, which is a mixture of cementitious materials and solid activators, similar to bagged commercial cement (Bayuaji et al., 2017). It not only optimizes the preparation process of geopolymer, but also guides the new development direction of geopolymer.

2. Materials and methods

2.1. Cementitious materials

The residual soil of fully weathered granite used in the experiment is widely distributed in southern China. It contains sand and gravel clay and is brownish red. The cohesive soil in the hard plastic state is a kind of construction dregs. The slag used in the test is S-95 grade granulated blast furnace slag powder, which is an industrial by-product. The samples are shown in Figure 1.

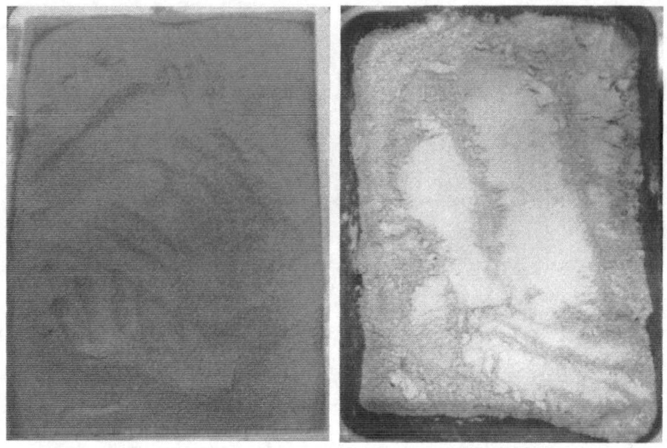

(a) CDG (b) Slag

Figure 1. Cementitious material used in the experiment.

The X-ray fluorescence analyses of the chemical composition of cementitious materials were carried out, and the results are shown in Table 1.

Table 1. Chemical composition of the cementitious materials

	SiO_2 (%)	Al_2O_3 (%)	Fe_2O_3 (%)	CaO (%)	MgO (%)	TiO_2 (%)	Na_2O (%)	K_2O (%)
CDG	45.71	36.92	8.96	0.49	-	4.40	-	2.96
Slag	31.57	15.27	0.23	43.18	6.68	0.74	0.45	0.21

2.2. Alkaline activator

The solid water glass is white granular crystal with a modulus of 3.28, wherein the SiO_2 content was 73.87% and the Na_2O content was 23.25%. Both calcium hydroxide and sodium carbonate are white powdery solids and are mainly used to adjust the modulus of the water glass solution. The samples of the alkaline activator are shown in Figure 2.

(a) Solid Water glass (b) Calcium hydroxide (c) Sodium carbonate

Figure 2. Alkaline activator used in the experiment.

3. Experimental methods

In order to simplify the preparation method of CDG geopolymer materials, the safety hazards in the preparation process are also reduced. The dry blending technique is used to premix the solid activator with the gelling material. In the preparation process of the construction site, such as ordinary concrete, it is only necessary to add the dry material and aggregate which are mixed in advance to the mixer, and then add water to stir to prepare the geopolymer material.

The solid activator and the gelling material were mixed and stirred for 60 s to uniformly mix the solid activator with the gelling material, and then stirred with water for 120 s. Then, the stirred slurry was poured into a test piece of 40 mm × 40 mm × 160 mm, and placed in a water mold for 1 d at room temperature and cured in water. The flexural strength and compressive strength of the specimens were measured for 28 d and 56 d. The measurement method is shown in Figure 3.

(a) Flexural strength (b) Compressive strength

Figure 3. Test piece strength measurement method

The determination of the flexural strength (R_f) requires a folding clamp, and the method of use is shown in Figure 3(a). The method for measuring the compressive strength (R_c) of the geopolymer paste is to put the half-cuboid after the fracture test of the fracture test into a compression clamp, so that the pressure receiving surface of the test piece is 40 mm×40 mm, and the use method is shown in Figure 3(b).

$$R_f = \frac{1.5 \times F_f l}{b^3} \quad (1)$$

$$R_c = \frac{F_c}{A} \quad (2)$$

In which, F_f is the load when the test piece is broken, l is the distance between the two supporting cylinders below the clamp, b is the length of the square section of the test piece, F_c is ultimate load when the compressed test piece is damaged, A is the pressure area of the test piece.

3.1. Mix ratio design of the activator modulus

The ratio of SiO_2/Na_2O in the activator was adjusted to 1.0, 1.2, 1.5, 2.0 by changing the ratio of solid water glass to $Ca(OH)_2$ and Na_2CO_3. The CDG and slag are mixed at a ratio of 3/2, the amount of Na_2O of the activator is controlled to be 6% of the cementitious material, the water-to-binder ratio is 0.5, and the blending thereof is shown in Table 2.

Table 2. Mix ratio design of the activator modulus

Label	CDG (g)	Slag (g)	$Ca(OH)_2$ (g)	Na_2CO_3 (g)	Na_2SiO_3 (g)	Water (g)
C1-1	900	600	78	111	120	750
C1-2	900	600	70	100	142	750

C1-3	900	600	60	87	177	750
C1-4	900	600	44	64	236	750

3.2. Mix ratio design of the activator dosage

Solid water glass, $Ca(OH)_2$ and Na_2CO_3 powder were used as solid activators, and the ratio was adjusted, therefore the ratio of SiO_2/Na_2O in the solid activator system was kept constant at 1.5. Adjust the proportion of cementitious materials, the total amount of residual soil of fully weathered granite is 60%, and the amount of slag is 40%. The amount of the activator was changed so that the Na_2O of the activator system was 4%, 6%, 8% of the cementitious material, and the water-to-binder ratio was 0.5. The specific mix ratio design is shown in Table 3.

Table 3. Mix ratio design of the activator dosage

Label	CDG (g)	Slag (g)	$Ca(OH)_2$ (g)	Na_2CO_3 (g)	Na_2SiO_3 (g)	Water (g)
C2-1	900	600	40	58	118	750
C2-2	900	600	60	87	117	750
C3-3	900	600	80	116	236	750

3.3. Mix ratio design of the water-to-binder ratio

The ratio of SiO_2/Na_2O in the activator was controlled to 1.5, and the residual soil of the fully weathered granite was mixed with the slag in a ratio of 3/2. The amount of the activator was 6% of the cementitious material calculated by Na_2O, and the water-to-binder ratio was changed to 0.4, 0.45, 0.5, 0.55, 0.6 respectively. The specific cooperation is shown in Table 4.

Table 4. Mix ratio design of the water-to-binder ratio

Label	CDG (g)	Slag (g)	$Ca(OH)_2$ (g)	Na_2CO_3 (g)	Na_2SiO_3 (g)	Water (g)
C3-1	900	600	60	87	177	600
C3-2	900	600	60	87	177	675
C3-3	900	600	60	87	177	750
C3-4	900	600	60	87	177	825
C3-5	900	600	60	87	177	900

4. Results and discussions

4.1. Influence of the activator modulus

The flexural strength and compressive strength of the test pieces at 28-day and 56-day during the curing age are shown in Figure 4.

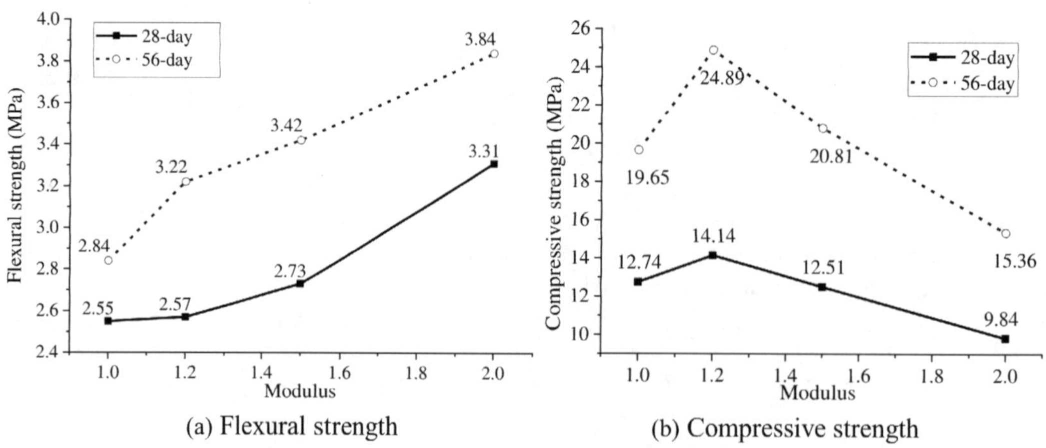

(a) Flexural strength (b) Compressive strength

Figure 4. Influence of the activator modulus on strength

It can be seen from Figure 4 that as the molar ratio of SiO_2/Na_2O in the activator increases from 1.0 to 2.0, the flexural strength of the cleaned specimen gradually increases; the compressive strength of the test piece first increases and then decreases. The maximum 56-day compressive strength of 24.90 MPa is obtained when the modulus is 1.2.

This can be attributed to the fact that the water glass with low modulus contains more oligosilicate tetrahedron, which is beneficial to the dissolution of silicon and aluminum components and promotes the geopolymer gel. However, due to the quick reaction, some micro-cracks are generated inside the test piece, resulting in a decrease in the bending strength. However, the compressive strength has no high flexural strength for microcrack sensitivity, so the compressive strength increases as the modulus decreases. When the ratio of Na_2CO_3 and $Ca(OH)_2$ is too high, too many microcracks leads to a decrease in strength.

4.2. Influence of the activator dosage

After the specimen is demolded, it should be cured in water to the predesigned curing time. Firstly, the flexural strength is measured, then the compressive strength of the two fractured half prism specimens is measured. Details of the strength versus activator dosage is shown in Figure 5.

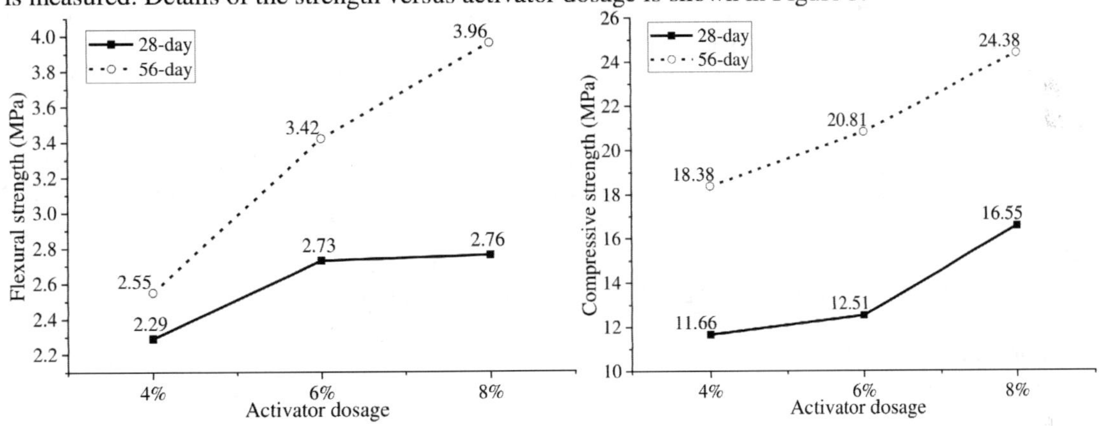

(a) Flexural strength (b) Compressive strength

Figure 5. Influence of the activator dosage on strength

It can be seen from Figure 5(a) that when the amount of Na_2O in the activator increased from 4% to 8% of the cementitious material, the flexural strength of the test piece increases. As the amount of activator increases, the increase of the flexural strength of the test piece from 28-day to 56-day is greater. Since the solid water glass has a low solubility in water under normal temperature and normal pressure conditions, it is required that the water glass solid is continuously dissolved in water to participate in the reaction. Therefore, the reaction is also slow. When the amount of the activator is large, the activator can always participate in the reaction, supplementing the microcracks in the test pieces, thus improving the bending strength.

From Figure 5(b), it can be found that the compressive strength of the test piece increases with the increase of the amount of the solid activator. When the amount of Na_2O is 8% of the cemented material, the compressive strength at 56-day is the largest, which can reach 24.38 MPa. This indicates that the amount of the activator has a greater influence on the mechanical properties of the test piece.

4.3. Influence of the water-to-binder ratio

Due to the low strength of the group 3-5 specimens during the 28-day curing period, the flexural strength and compressive strength of the 56-day were not studied. Therefore, there were only four specimens for the 56-day test. The test result data is shown in Figure 6.

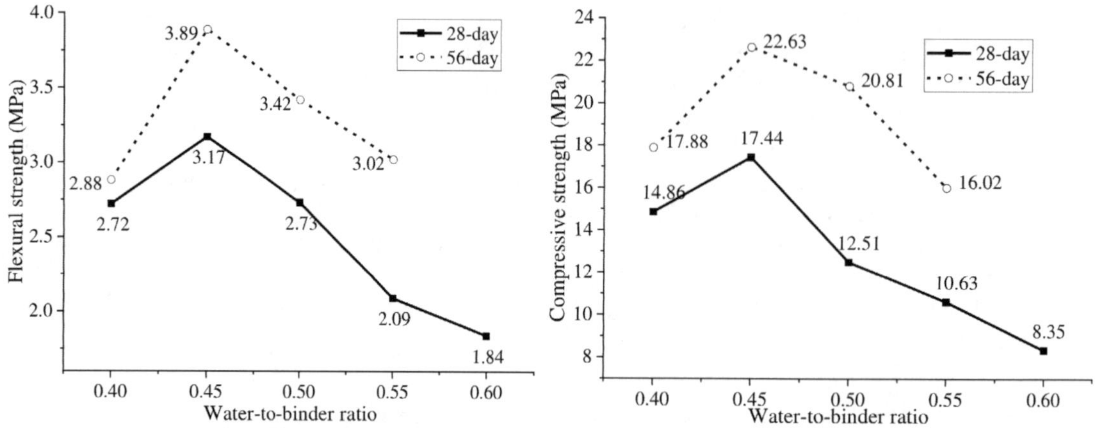

(a) Flexural strength (b) Compressive strength

Figure 6. Influence of the water-to-binder ratio on strength

It can be seen from Figure 6 that, as the water-to-binder ratio increases, the flexural strength and compressive strength of the test piece increase first and then decrease, the test piece with the water-to-binder ratio equals to 0.45 can obtain the highest strength. The flexural strength of the 56-day is 3.90 MPa and the compressive strength is 22.63 MPa.

When the water-to-binder ratio is relatively high, a large amount of moisture in the test piece remains inside the test piece, forming capillary pores, which reduces the compactness of the test piece, resulting in a decrease in the strength of the test piece. When the water-to-binder ratio is too low, the fluidity of the geopolymer material is poor, the test piece is difficult to form, and the defects in the pouring process would affect the strength of the geopolymer test piece.

5. Conclusion

In order to simplify the preparation process of geopolymer materials, further promoting the development of construction materials, dry-mixed geopolymer materials are prepared through the dry mixing method. Optimizing the geopolymer preparation process can solve the difficulties in transporting alkaline activator solutions.

In this research, a series of studies on the mechanical properties of dry-mixed geopolymer materials were carried out. The effects of the activator modulus, the activator dosage, and the water-to-binder ratio on the mechanical properties were investigated. For the dry-mixed geopolymer paste material, when the molar ratio of the SiO_2/Na_2O to the initiator is 1.2, the compressive strength is the highest. In addition, the flexural strength increases with the increase of the modulus. As the amount of activator increases, the strength of the test piece also increases. When the water-to-binder ratio varies from 0.4 to 0.6, the strength of the geopolymer paste first increases and then decreases. When the water-to-binder ratio is 0.45, the highest 56 d compressive strength 22.63 MP is achieved. Additionally, during the 56-day curing period, the dry-mixed geopolymer material still has a certain increase in strength.

Acknowledgement

This research is funded by the National Nature Science Foundation of China (Grant No. 51578181), Natural Science Foundation of Guangdong Province (No. 2016A030313665) and Shenzhen Science and Technology Plan Project (No. JCYJ20150327155221857). Acknowledgement is also given to the Shenzhen Carbon Storage Cement-Based Materials Engineering Laboratory.

References

[1] Davidovits J. Global warming impact on the cement and aggregates industries[J]. World Resource Review; (United States), 1994, 6:2.

[2] Shi H, Xia M, Guo X. Research Development on Mechanism of Fly Ash-based Geopolymer and

Effect of Each Component[J]. Journal of the Chinese Ceramic Society, 2013, 41(7):972-980(9).

[3] Zha X, Zhu J, Experimental Study on the Effects of Calcium-rich Materials on the Early Compressive Strength of Completely Decomposed Granite (CDG) Base-geopolymer[C]// 4th International Symposium on Hydrogen Energy, Renewable Energy and Materials, HEREM 2018, 2018:2-8.

[4] Suwan T, Fan M. Influence of OPC Replacement and Manufacturing Procedures on the Properties of Self-Cured Geopolymer[J]. Construction & Building Materials, 2014, 73(1-2):551-561.

[5] Suwan T, Fan M, Braimah N. Internal Heat Liberation and Strength Development of Self-Cured Geopolymers in Ambient Curing Conditions[J]. Construction & Building Materials, 2016, 114:297-306.

[6] Bayuaji R, Yasin A K, Susanto T E, et al. A Review in Geopolymer Binder with Dry Mixing Method (Geopolymer Cement)[C]// American Institute of Physics Conference Series. American Institute of Physics Conference Series, 2017:020-022.

Recovery of valuable metal from Photovoltaic solar cells through extraction

Wei-Sheng Chen[1], Yen-Jung Chen[1*], Kai-Chieh Yueh[1], Chii-Pwu Cheng[2], Tien-Chin Chang[2]

[1] Department of Resource Engineering, National Cheng-Kung University, Tainan, R.O.C.

[2] Institute of Environmental Engineering and Management, National Taipei University of Technology, Taipei, R.O.C.

*E-mail: qaz5932201@gmail.com

Abstract. The installation of PV modules was 97.9GW and the accumulation volume of PV device was 500GW in 2018 According to the research, the accumulation of waste modules will reach to 8600 tons in 2030 as the result of the life expectancy of PV modules. Moreover, Crystalline-Silicon solar panels account for 90% of the waste. This study recycles photovoltaic solar cells by leaching and extraction. According to the analyst, Silicon cells content 90% of Si, 0.7% of Ag, and 9.3% of Al. Silicon cells were leached by 4M nitric acid at 80°C for 4 hours then 3M sodium hydroxide at 70°C for 3 hours, and the leaching efficiency were 99.7% of Ag, and 99.9% of Al, respectively. Leaching process separated Silicon from others metal. Na-Cyanex 272 in kerosene was employed to separate Al from Ag. The efficiency of extraction is 96%. After the process of extraction, 1M hydrochloride acid was employed to strip Al from the organic solvent. The efficiency of stripping is 99.9%. After extraction process, Ag was remained in the aqueous phase and precipitated to AgCl. In conclusion, this study shows the optimal parameters of the leaching steps, extraction steps, precipitation and vaccum concentration to get the product. The recovery of Si and Ag were 99.5% and 98% respectively.

Keywords: Photovoltaic modules, PV cells, Hydrometallurgy, extraction, recycle

1. Introduction

PV modules is one of the greenest and promoted energy-generator. By the statistic, 106GW of solar PV capacity added in 2018 and the accumulation of PV modules is now increased to 508GW.[1] More solar PV was installed than the net capacity additions of fossil fuels and nuclear power combined. PV modules have an average life of 20-25 years [2], which means that at 2030 we will have the great amount of PV waste in the future. According to the statistic in IRENA, the waste of solar panels in 2050 will reach 600,000 tons, which means that we should make effort on recycle PV modules right now.

According to the experiment, Poly-crystalline Silicon solar panels contain Silicon, Aluminum, silver, Lead, and Tin and PV cells contain the most of the Silicon (90%), Aluminum and Silver (1%). This study dismantled PV modules to PV cells and recovered PV cells by leaching and extraction process. Silicon is hard to leach by non-organic acids (HCl, H2SO4, and HNO3). The first step to purify silicon was leaching metals from PV cells and purify silicon. Aluminum is known as easy to separate by precipitate and extraction. This study decided to use extraction process to separate Aluminum from silver to improve the purity of silver. According to the literature review [4]-[12], N503, Tributyl Phosphate (TBP), D2EHPA (di-(2-ethylhexyl) phosphoric acid), Cyanex901 and Cyanex272 was used to separate Aluminum from other metals. This study used Cyanex272 for extractant because of the optical pH value 2-3. Cyanex272 and D2EHPA was in the pH range 2-3 for effectively extract Aluminum(III). Therefore, Cyanex272 has higher efficiency and selectivity for Aluminum(III) which

IOP Conf. Series: Materials Science and Engineering **720** (2020) 012007 doi:10.1088/1757-899X/720/1/012007

makes Cyanex272 the better agent for selectively extraction.[13] After extraction and stripping process, precipitation was employed to recover Silver to AgCl. [14][15]

This study provided the separation and recovery of Silicon, Aluminum and silver from PV cell via hydrometallurgy to obtain valuable metals. The resources can be recycled and also reduced the wastes of PV modules.

2. Materials and experiment

2.1 Materials

Poly-crystalline silicon PV modules used in this study are from the waste PV modules recycling factory. Table 1 shows the mass fraction of PV modules. PV modules are composed of aluminum frame, tempered glass, Ethylene-vinyl acetate resin and back sheet, PV cells, Ribbon, and Junction Box. This study dismantled PV moduels to PV cells and recovered it by several processes. Table.2 shows the chemical composition of commercial PV cells.

Table 1. Mass fraction of PV modules

Aluminum frame	Tempered glass	EVA and backsheet	PV cells	Ribbon
14.7%	68.4%	11.2%	3.1%	1.3%

Table 2. The chemical composition of PV cells

Element	Ag	Al	Si
Mass fraction	0.7%	9%	90%

2.2 Pretreatment

PV cells was dismantled from the PV modules. After removing Aluminum frame and tempered glass, EVA reins still remained on the PV cells. Pretreatment process was employed to eliminate the EVA resin from the PV cells.

2.3 Leaching

Nitric acid was employed to leach Al and Ag from the PV cells. After acid leaching process, Al-Si alloy still remained on the PV cells. In order to purify the Silicon, Sodium Hydroxide leaching was employed to break the boning between Si and Al. To optimize the experimental condition for effective leaching, this study was made carrying different process parameters viz. time (0.5-4hr), liquid-solid ratio, and acid concentration (0.5M-6M).

Leaching efficiency of metal was calculated by the equation below:

$$X\% = \left(\frac{V_c C_x}{M w_x}\right) \times 100\% \quad (1)$$

X%=leaching efficiency(g/L), M=the weight of the sample(g), Vc=the volume of the liquid(L), Wx=target metals weight fraction (wt%).

2.4 Extraction

The metals in the leaching solution is separated through selectively solvent extraction and stripping. Aluminum was extracted by Na-Cyanex272 and silver remained in the solution. Using HCl to strip Aluminum from the organic phase. The metals can be separated. The solution with silver was precipitate by sodium chloride into AgCl. This study also carried the parameters of extraction process viz. the necessary of saponification, pH value (0.5-3), O/A ratio, extractant concentration, stripped solvents concentration, and reaction times.

The extraction efficiency of Na-Cyanex272 and stripping efficiency were calculated by equations below:

$$E\% = \left(\frac{\Sigma[M]_{org}}{\Sigma[M]_{org} + \Sigma[M]_{aq}}\right) \times 100\% \ (2)$$

$$E_{stripping}\% = \left(\frac{\Sigma[M]_{after\ stipping}}{\Sigma[M]_{before\ stipping}}\right) \times 100\% \ (3)$$

$$Na^{+}_{(aq)} + H^{+}\ Cyanex272^{-}_{(aq)} \rightarrow Na^{+}_{(aq)}\ Cyanex272^{-}_{(aq)} + H^{+} \ (4)$$

2.5 Analytical method

The sample after leaching, extraction process, and precipitation were filtered by 0.45 μ m of membrane filter and were diluted by 1% HNO3 solution for ICP-OSE (Inductively Coupled Plasma-Optical Emission Spectrometry, PerkinElmer optima 2100DV) analyst. The ICP-OSE analyst was calibrated with ICP multi-element stander and tin standard solutions. Each aqueous solution was analyzed for three times and averaged for the reported.

3. Result and discussion

3.1 Thermal treatment to eliminate EVA resin from PV cells

After dismanlting PV modules, EVA resin still remained on the PV cells which might harm to leaching and the process below. Fig. 1 (a) and (b) show the TG/DTA analyst chart of PV cells with EVA resin and pure EVA reasin. Comparing to figures, both of them decline the weight on 300 degree. After heating to 500 degree, the mass changing tend to balance. This study used the thermal process to eliminate EVA resin by heating to 500 Celsius degrees for 5 hours in the atmosphere. This process can eliminate 99.97% of EVA resin from PV cells.

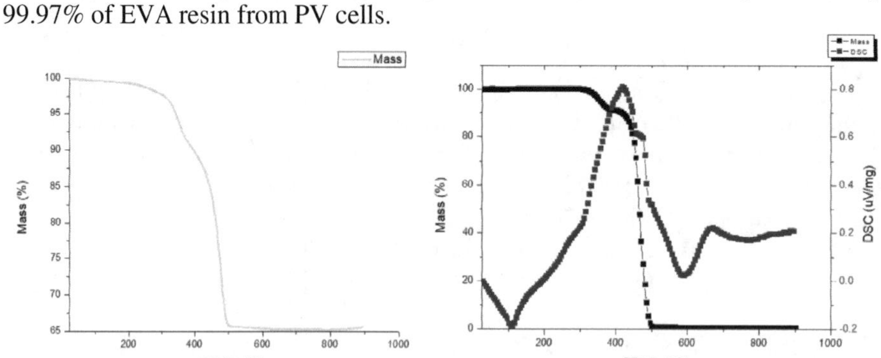

Fig. 1 (a) PV cells with EVA resin layers TG analyst. (b) TG analyst of EVA resin.

3.2 Nitric acid for first leaching process

This study provides the leaching process to leach Ag and Al form PV cells through Nitric acid. Nitric acid can effectively separate Ag and 70% weight of Al from PV cells.

3.2.1 Effect of the concentration of nitric acid

IOP Conf. Series: Materials Science and Engineering **720** (2020) 012007 doi:10.1088/1757-899X/720/1/012007

Fig.3 shows the leaching efficiency from adjusting the concentration of HNO4 in 4 hours, 50 of L/S ratio, and 85 degrees of temperature. Silver and aluminum both get a good leaching efficiency on 4M of nitric acid. After rising the concentration of HNO4, the leaching efficiency of aluminum still not overcome 75%. Al-Si alloy still remained on the PV cells. The optimal condition of the concentration was chosen as 4M.

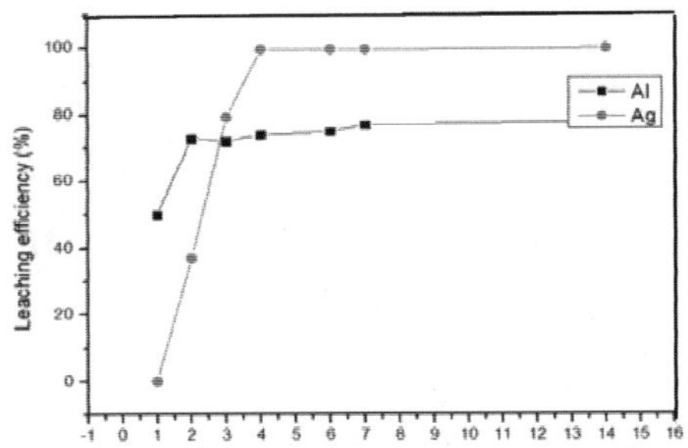

Fig. 2 Leaching efficiency of adjusting [HNO4] (L/S=50, Reaction times=240min., Temperature=85 degree).

3.2.2 Effect of Liquid-solid ratio

Fig.4 show the leaching efficiency from adjusting the liquid-solid ratio. The L/S ratio was investigated by verifying the ratio from 10 to 400. As the result, PV cells has effective leaching efficiency (99.5% of silver and 70% of aluminum) after 100. Hence, the optimal condition of the L/S ratio was chosen as 100ml/g.

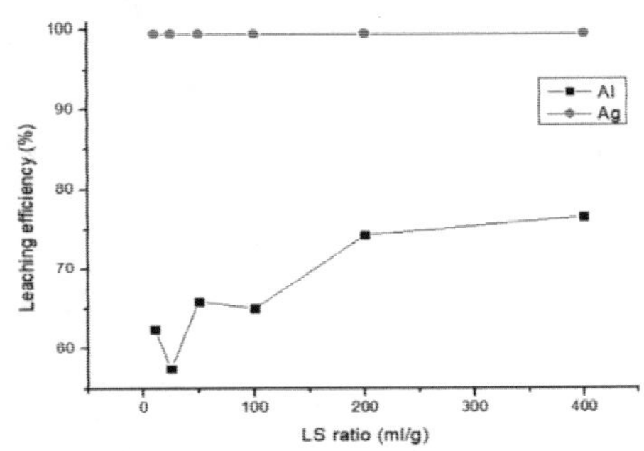

Fig. 3 Leaching efficiency of adjusting L/S ([HNO4]=4M, reaction times=240min., Temperature=85 degree).

3.2.3 Effect of the reaction times

Fig. 5 shows the leaching efficiency from adjusting the reaction times for 4M of HNO4 in 100 of LS ratio, and 85 degrees. Reaction times was investigated by verifying from 30 minutes to 240 minutes. As the result, after and hours, both of the leaching efficiency tend to balance. Hence, the optimal condition of reaction times was chosen as 60 minutes.

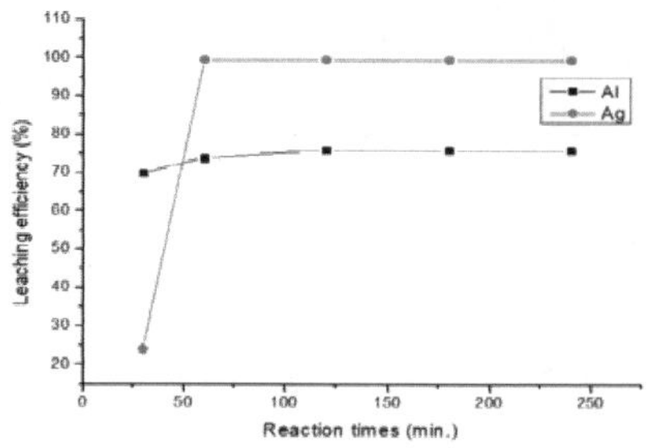

Fig. 4 Leaching efficiency of adjusting reaction times ([HNO4]=4M, L/S=100, Temperature=85 degree).

3.3 Sodium Hydroxide leaching process

After nitric acid leaching process, Al-Si alloy still remained on the PV cells. As the equation showing below:

$$\text{AlSi} + 2\text{NaOH} + H_2\text{O} \rightarrow 2\text{Na}^+ + Si_2O_3 + Al^{3+} + H_2 \quad (5)$$

Sodium Hydroxide has the ability to transfer Al to Al ion. This study set the optimal parameters as 3M of NaOH, L/S ratio=50, 240minutes, and temperature for 70 degrees. After two leaching processes, the leaching efficiency of Al ion can reach to 99.9%.

3.4 Alumnium extraction through Na-Cyanex272

After leaching process, Na-Cyanex272 was employed to extract Aluminum from Silver. Na-Cyanex272 was diluted by kerosene. Cyanex272 was saponificated by sodium hydroxide into Na-Cyanex272. As Fig. 6 and Fig. 7 show, the extraction efficiency of aluminum with saponification is higher than non-saponification.

3.4.1 Effect of pH value

The pH value was adjusted with nitric acid and ammonia solutions. The pH values were verified from 1 to 4. After the pH value reach 4.2, aluminum started to precipitation into Al(OH)3. According to Fig. 6, the extraction efficiency increased at 2.5. The extraction efficiency was 96% at pH 3.7. Hence, this study set 3.7 as the optimal parameters.

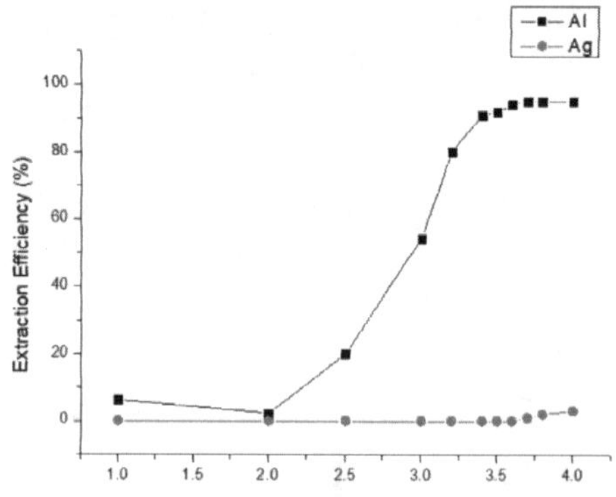

Fig. 5 The extraction efficiency of Na-Cyanex272 in condition of 0.3M extractant, A/O ratio=1, and reaction times=10min.

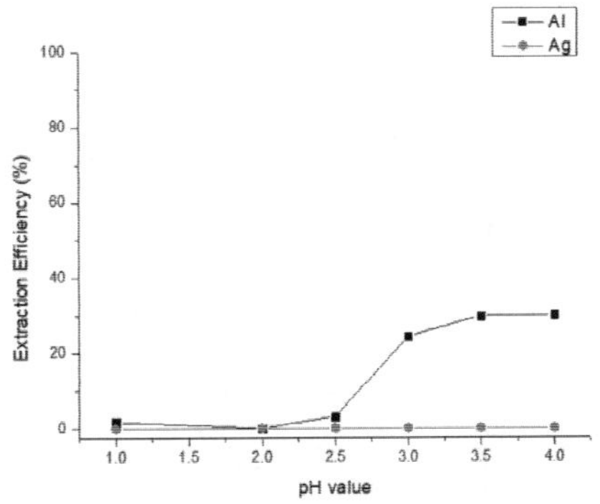

Fig. 6 The extraction efficiency of Cyanex272 in condition of 0.3M extractant, A/O ratio=1, and reaction times=10min.

3.4.2 Effect of Na-Cyanex272 concentration

The concentration of Na-Cyanex272 was verified from 0.1M to 0.6M. Fig. 8 shows the extraction efficiency of the concentration of Na-Cyanex272. The extraction efficiency increased from 3% to 96% when the concentration of extractant increased from 0.1M-0.6M. After reaching the concentration of extractant to 0.3M, the efficiency tend to balance. Hence, the optimal parameters of the extractant concentration was set as 0.3M.

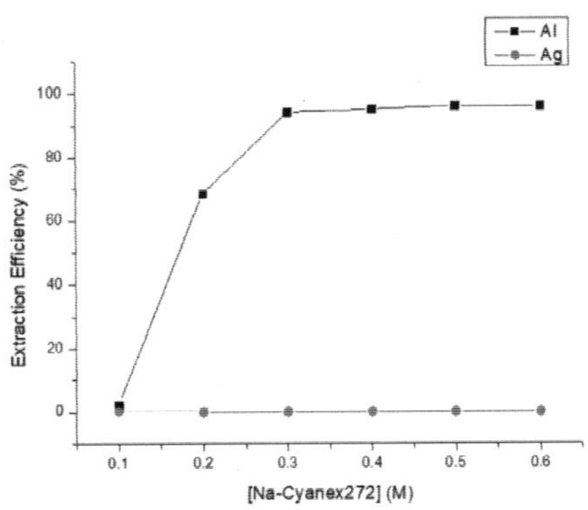

Fig. 7 Extraction efficiency of adjusting the concentration of Na-Cyanex272

3.4.3 Effect of extraction aqueous-oil volume ratio

The effect of A/P ratio was investigated and set from 0.1 to 10. Fig. 9 was the extraction efficiency which decreased from 97% to 26.8%. In order to concentrated aluminum in high concentration and reduced the waste of extractant. This study set the optimal parameters of A/O ratio as 3(mL/mL).

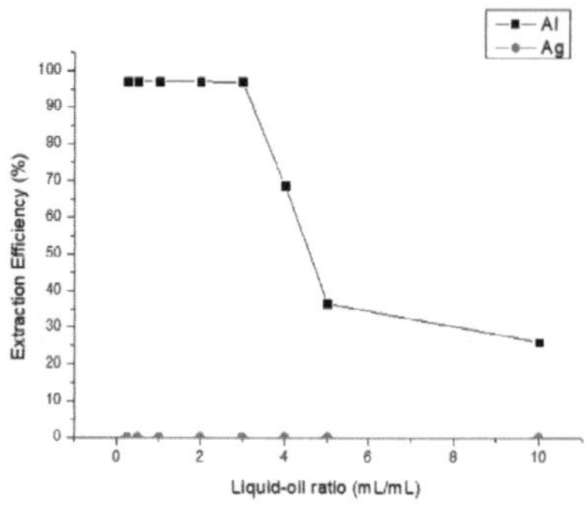

Fig. 8 Extraction efficiency of adjusting aqueous-oil volume ratio

3.4.4 Effect of adjusting reaction times

Under the condition of fixed solution pH value=3.7, A/O=3, and [Na-Cyanex272]=0.3M, the shaking time was verified to 1min., 5min., 10min., 15min., 20min., and 30min., individually. Fig. 10 shows the result of verified the reaction times. After the time reach 5min., the efficiency tends to balance. Hence, this study set 5min as the optimal parameters of reaction times.

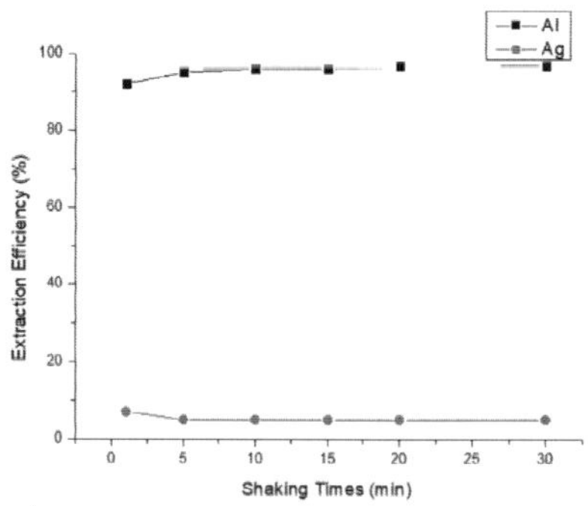

Fig. 9 Extraction efficiency of adjusting reaction times

3.4.5 Stripping experiment

From the extraction experiments, it was observed that Na-Cyanex272 has the ability to extract aluminum from silver-aluminum solution. The organic phase with aluminum was stripped to water phase by different concentration of hydrochloric acid, OA ratio and reaction times. The results indicated that 1M HCl, O/A ratio=0.25, and reaction times=5 min. can efficienctly strip the aluminum from organic phase. The efficiency of stripping was 99.9%.

Table 3. Optimal condition of Al3+ stripping

Optimal condition of Al^{3+} stripping	
Reagent	1M HCl
O/A ratio	0.25
Reaction times	5 min
Stripping efficiency	99.9%

IOP Conf. Series: Materials Science and Engineering **720** (2020) 012007 doi:10.1088/1757-899X/720/1/012007

3.5 Silver chloride and purified silicon after leaching and extraction processes

After leaching and extraction processes, silicon, aluminum, and silver were separated effectively. Silicon was collected after leaching process due to the selectivity of nitric acid. The composition after leaching process was showed on Table. According to the reaction equation below:

$$Ag^{+}_{(aq)} + Cl^{-}_{(aq)} \rightarrow AgCl_{(s)}$$

The standard Gibbs free energies of Ag+, Cl- and AgCl(s) are 77.16kJ/mol, -131.0563kJ/mol, and -109.86kJ/mol, respectively. The solubility product (Ksp) of AgCl is 10-9.82 at 25°C.[16] This reason indicated Ag ion precipitated easily and rapidly as AgCl then separated with Al ions. Hence, this study precipitated Ag ion into silver chloride for recovering Silver in the PV cells. The recovery of Si and Ag were 99.1 and 95% individually. Therefore, the purity of silicon after the processes is 99.5%.

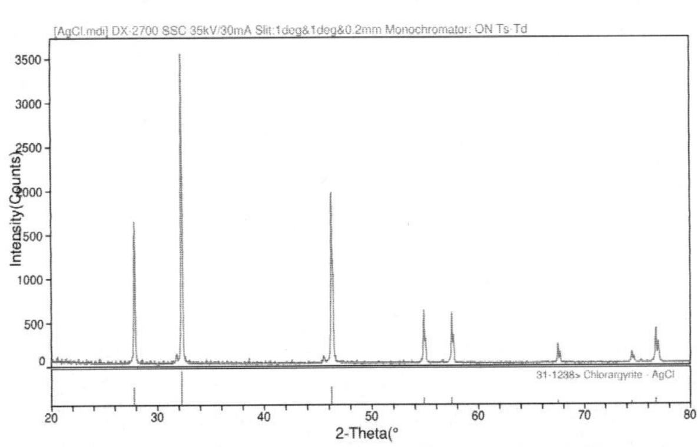

Fig. 10 Silver Chloride after precipitated by Sodium Chloride

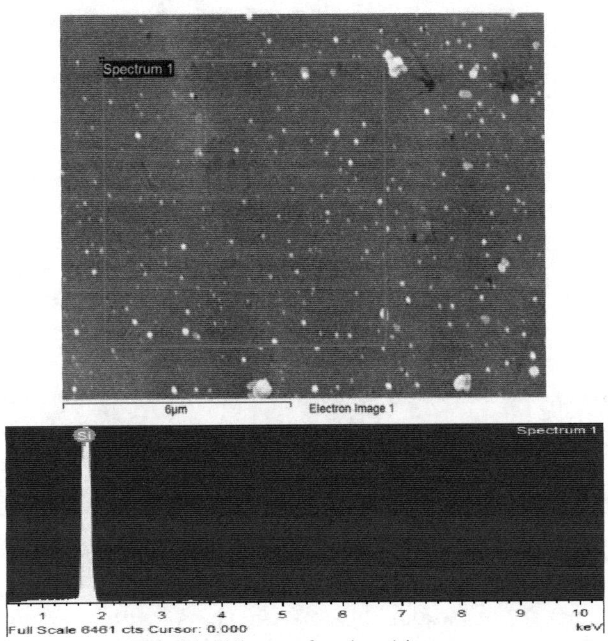

Fig.11 Silicon after leaching process

Table 4. Composition of silicon wafer before and after leaching process (Analyst by ICP-OSE)

Element	Before leaching	After leaching
Ag (mg/kg)	5701.8	22.8
Al (mg/kg)	78802.8	101.4
Na (mg/kg)	1169.4	3850
B (mg/kg)	75.77	74
Cu (mg/kg)	2.3	>0.01
Pb (mg/kg)	6.63	6.95
Si (purity %)	90%	99.5%

4. Conclusion

This study proposed a recovery route of silicon and silver from PV cells by hydrometallurgy ways. Two leaching steps (acid and Alkaline) were employed to purify silicon and the purity of silicon was 99.5%. The leaching rate of Al and Ag were up to 99% after two leaching process. Moreover, the recovery of Ag and Si was 98% and 99.5.%. This study can provide to the research of recycling PV modules.

5. Acknowledge

This study was supported by National Cheng-Kung University Dept. of Resource Engineering, R.O.C.

References

[1] T. Doi, I. Tsuda, H.Unagida, A.Murata, K.Sakuta, and K.Kurokawa, "Experimental study on PV module recycling with organic solvent method," Sol. Energy Mater. Sol. Cells, vol. 67, no. 1–4, pp. 397–403, 2001.

[2] S.Weckend, A.Wade, and G.Heath, End-of-life management: Solar photovoltaic panels. 2016.

[3] Y. K.Yi, H. S.Kim, T.Tran, S. K.Hong, andM. J.Kim, "Recovering valuable metals from recycled photovoltaic modules Recovering valuable metals from recycled photovoltaic modules," *J. Air Waste Manage. Assoc.*, vol. 64, no. 7, pp. 797–807, 2014.

[4] T.N. Shilimkar, S.S. Kolekar, M.A. Anuse, Rapid extraction separation of aluminium(III) from associated elements with n-octylaniline from succinate media, Sep. Purif. Technol. 42 (2005) 55–63.

[5] W. Hiroshi, K.Yasushi, Extraction of trivalent aluminium and gallium from sodium hydroxide solution containing tartaric acid by trioctylmethylammonium chloride, Solvent Extr. Res. Dev. Jpn. 7 (2000) 118.

[6] I. Toth, E. Brucher, Z. Szabo, Extraction of gallium(III) and aluminium(III) with o,o-dialkyldithiophosphoric acids, Talanta 37 (1990) 1175–1178.

[7] P.E. Tsakiridis, S. Agatzini-Leonardou, Solvent extraction of aluminium in the presence of cobalt, nickel and magnesium from sulphate solutions by Cyanex 272, Hydrometallurgy 80 (2005) 90–97.

[8] M.M. Orive, M.A. Olazabal, L.A. Fernandez, J.M. Madariaga, The recovery of cobalt and nickel from acidic sulphate solutions in the presence of aluminium, Solvent Extr. Ion Exch. 10 (1992) 787–797.

[9] H.S. Ajgaonkar, P.M. Dhadke, Solvent extraction separation of iron(III) and aluminium(III) from other elements with Cyanex 302, Talanta 44 (1997)563–570.

[10] Cytec, Cyanex 272, Technical Brochure, American Cyanamid Company, 1989.
[11] K. Sarangi, B.R. Reddy, R.P. Das, Extraction studies of cobalt(II) and nickel(II) from chloride solutions using Na-Cyanex 272. Separation of Co(II)/Ni(II) by the sodium salts of D2EHPA, PC88A and Cyanex 272 and their mixtures, Hydrometallurgy 52 (1999) 253–265.
[12] G.M. Ritcey, A.W. Ashbrook, Solvent Extraction Principles and Applications to Process Metallurgy, Part-1, Elsevier Science Publishers B.V., Amsterdam, The Netherlands, 1984, pp. 15–20.
[13] D.Mohapatra, K.Hong-In, C. W.Nam, andK. H.Park, "Liquid-liquid extraction of aluminium(III) from mixed sulphate solutions using sodium salts of Cyanex 272 and D2EHPA," *Sep. Purif. Technol.*, vol. 56, no. 3, pp. 311–318, 2007.
[14] P.Dias, S.Javimczik, M.Benevit, H.Veit, andA. M.Bernardes, "Recycling WEEE: Extraction and concentration of silver from waste crystalline silicon photovoltaic modules," Waste Manag., vol. 57, pp. 220–225, 2016.
[15] I. F. F.Neto andH. M. V. M.Soares, "Sequential separation of Ag , Al , Cu and Pb from a multi-metal leached solution using a zero waste technology," *Sep. Sci. Technol.*, vol. 53, no. 18, pp. 2961–2970, 2018.
[16] Soolyung Kim, J.C. Lee, K.S. Lee, K.K. Yoo, and R.D. Alorro, 2014: Separation of Tin, Silver and Copper from Waste Pb-free Solder Isong Hydrochloric Acid Leaching with Hydrogen Peroxide, Materials Transactions, Vol. 55, No. 12, 1885-1889

IOP Conf. Series: Materials Science and Engineering **720** (2020) 012008 doi:10.1088/1757-899X/720/1/012008

Cobalt−Citrate Metal−Organic-Framework UTSA-16 on TiO_2 Nanoparticles

Mei-Yi Tseng[1], Yen-Hsun Su[2], Yi-Sheng Lai[3], Fei Pan[4], Po-Yen Kung[5]

[1, 2, 3,5] Department of Materials Science and Engineering, National Cheng Kung University

[3]Institute of Physics, University of Münster, Wilhelm-Klemm-Strasse 10, Münster 48149, Germany

[4]Physics Department, Technische Universitat Mu¨nchen, James-Franck-Straße 1, Garching 85748, Germany

[4]Laboratory for Biointerfaces, Swiss Federal Laboratories for Materials Science and Technology (EMPA), Lerchenfeldstrasse 5, St. Gallen 9014, Switzerland

louisaq2014@gmail.com,yhsu@mail.ncku.edu.tw, zkexmoon@gmail.com, Fei.Pan@empa.ch, poyen0510@gmail.com

Abstract. Over the past few decades, the multifunctional material, metal-organic frameworks(MOFs), has been distinguished for their unique properties such as reticular structure, permanent porosities, high surface area, straightforwardly self-assembly and crystalline nature. On a count of these potentialities, the applications of MOFs can be classified into separation, catalysis, sensing, and biomedicine these four fields. In our studies, we will concentrate on the fields of gas storage and separations, water splitting, CO_2 reduction, and White-light emission and so on.

Keywords: metal-organic frameworks(MOFs), UTSA-16, carbon dioxide capture, titanium dioxide, photocatalysis

1. Introduction

Nowadays, one of the urgent global problems for us is the wide use of fossil fuels and environmental pollution. These troubles not only damage human bodies but also rise some energy crises for human beings. For instance, carbon dioxide emissions which have grown dramatically ever recorded in recent decades increase the greenhouse effect[1]. These emissions are primarily resulted from the combustion of fossil fuel, natural gas and coal, which are always from our daily life especially for the developing counties. The industrial development and economic growth usually come with environmental concerns that we are supposed to be confronted with the issue of civilization and globalization seriously[2]. Therefore, the development of the methodology how to capture and separate carbon dioxide has been extremely significant in order to decrease the environmental impact.

As for the part of capturing carbon dioxide, aqueous alkanol-amine absorbents are very useful to the chemical absorption. Nevertheless, this current technology has some weaknesses during the process of capture as followings. It is necessary to need a higher energy requirement for the regeneration of the solvent for the removal of acidic gases such as carbon dioxide[3]. And it is the reason that carbon dioxide can be released only at a high raising temperature due to the strong interactions between carbon dioxide and these absorbents. In addition, the high boiling points of these alkanol-amine absorbents is another reason why this kind of technology cost an enormous number of energy[4]. What's worse, carbon dioxide capture from a post-combustion gas through this technology leads to an energy penalty. Because of the corrosive natural property of the absorbents, they bring the hazard damage to the environment as well.

Among these feasible materials, metal-organic frameworks (MOFs) have emerged as a phenomenal carbon dioxide adsorbent. Metal-organic frameworks are crystalline materials that consist of coordination bonds between transition-metal cations and multidentate organic linkers through strong metal-ligand coordination bonds. The structure of MOFs is characterized by an open framework that can be porous. In addition, due to their unique properties such as reticular structure, crystalline nature, controllable structure, permanent porosities, high surface area, straightforward self-assembly, metal-organic frameworks has attracted increasing attention in the clean energy applications[5]. In particular, the properties of permanent porosities and high surface area, such a physical adsorption provides an alternative way to capture and separate carbon dioxide very easily. According to the previous document, the metal-organic frameworks, UTSA-16, which empirical formula is

[K(H2O)2Co3(cit)(Hcit)] (with cit = fully dehydroxylated citrate anion (4-) and Hcit = citrate anion (3-)), displays high uptake (160 cm3 cm^{-3}) of CO_2 at ambient conditions[6]. This MOF also has high separation selectivity and capacity for CO_2 capture applications.

It is very essential to mention that the cobalt–citrate metal–organic framework (UTSA-16) has the property of oxygen-evolution which is able to efficiently electro-catalyzing water oxidation[7]. UTSA-16, which we concentrate on in this study, has a three-dimensional anatase-type structure. It was composed of tetrahedral Co(II) atoms linked to tetranuclear cobalt citrate clusters in octahedral geometry. Each potassium ion coordinates two water molecules. Around them, there are two hydroxyl groups belonging to Carboxylic acid groups of the citrate units, which would be present conferring to the site a particularly interesting reactivity. [8]

On the basic of the previous popular paper[9], we can further explore more applications because of the multifunctional MOF materials beyond the traditional organic/inorganic hybrid materials. The multifunctional MOF materials have extremely extensive applications, i.e. gas storage and gas separation, water splitting, carbon dioxide reduction, white-light emission, chemical sensors and so on. And in other word, metal-organic frameworks could be applied in many fields like optical, electric and magnetic materials, catalysis, or biomedicine. In our research, we tend to focus on whether UTSA-16 synthesized with TiO_2 nanoparticles has the dual effect of carbon dioxide capture and water splitting, and the interaction between the MOF and TiO_2 nanoparticles. And thanks to the characteristic of high stability, low cost and non-toxic nature, TiO_2 materials are one of the most widely photocatalysts to date[10]. Moreover, we would synthesize a variety of concentrations of Cobalt–citrate metal–organic framework in the experimental section below so as to search the optimum effect within those composites. Since our research is still in progress, in the following discussion I'll refer to some papers as a comparison.

2. Materials and experiments

2.1. Materials

Cobalt acetate ($Co(Ac)_2 \cdot 4H_2O$), citric acid($C_6H_8O_7 \cdot H_2O$), ethanol, deionized water, potassium hydroxide (KOH) and ITO coated glass were used as received from Uni-onward corporation.

2.2. Synthesis of UTSA-16

Cobalt–citrate metal–organic framework (UTSA-16) was prepared by a solvothermal method in the light of the previous papers. [7, 8] First, $Co(OAc)_2 \cdot 4H_2O$ (0.249 g, 1 mmol) were dissolved in 10 mL of a mixed ethanol/water (1/1 v/v) solution in the 45 mL Teflon-lined autoclave. Then we added $C_6H_8O_7 \cdot H_2O$ (0.21 g, 1 mmol) 、KOH (0.168 g, 3 mmol)slowly and sequentially. And the mixed solution should be magnetically stir for at least 10 min when adding another new material. After mixing for a period of time, the Teflon-lined autoclave was sealed in a hydrothermal kettle and was heated to 120 °C in a furnace for 24 h. After that, the autoclave was naturally cooled to room temperature. The violet precipitates were collected by centrifugation and washed with deionized water and ethanol for three consecutive times. Finally, the violet powder was dried at 60 °C for 2~3 h in a furnace. In the beginning of experiment, the color of cobalt acetate diluted in in ethanol and DI water was magenta. However, UTSA-16 transformed to violet after heating for 24 hours. Herein, the UTSA-16 could also be synthesized by transforming the total volume of ethanol/ deionized water (1/1 v/v) to 7.5/7.5, 10/10, 15/15 and 20/20 mL respectively.

2.3 Synthesis of UTSA-16/ ITO or TiO_2 composites

As for the synthesis of UTSA-16/ ITO nanocomposites, we employed the preceding approach of UTSA-16 to the development of UTSA-16 on the ITO coated glass. That is, the synthesis was established on the pure UTSA-16. After we finished the process of mixing $Co(OAc)_2 \cdot 4H_2O$ (0.249 g, 1 mmol) 、$C_6H_8O_7 \cdot H_2O$ (0.21 g, 1 mmol) 、KOH (0.168 g, 3 mmol) 、ethanol/ deionized water, two pieces of 1cm x 2cm ITO coated glass were cautiously placed at the bottom of the Teflon-lined autoclave. It is important to mention here that the conductive slide must face upward. The latter procedures of heating and collecting sample were the same as synthesis of UTSA-16. The synthesis of UTSA-16/ TiO_2 nanocomposites was definitely similar to the synthesis of UTSA-16/ ITO composites. The difference was that the substrate was substituted for TiO_2 glass. All of the materials needed, and steps were equal. The mixed solution was heated at 120 °C for 24 h before placing two pieces of 1cm x 1cm TiO_2 glass. (Figure.1)

2.4 *Device fabrication*

ITO coated glasses were first cut into two pieces from the original size, and then cleaned using soap, ethanol and deionized water for the purpose of avoiding contamination. It's more convenient and safe to conserve these ITO coated glasses in isopropanol. Usually, the multimeter was used to measure the conductive slide of ITO coated glasses.

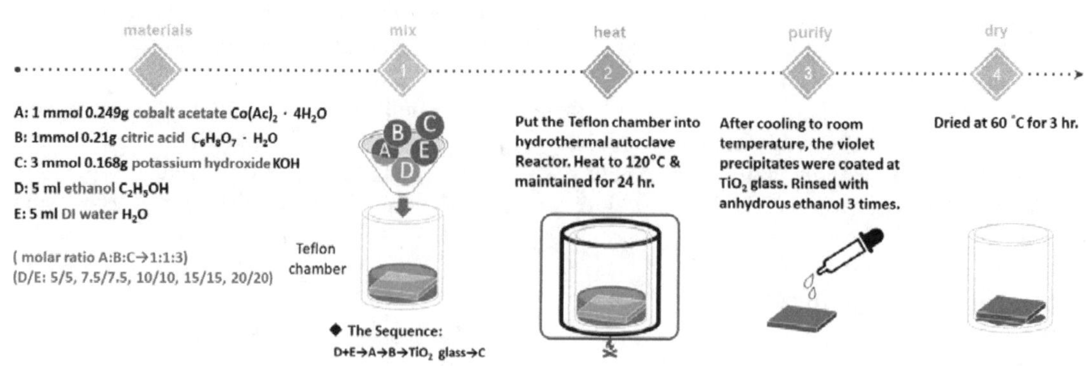

Figure1 synthesis of UTSA-16 on TiO_2 nanocomposites

3. Result and discussion

In our experimental section, because ITO glass is one of the most widely used and most popular transparent conducting oxides, we make an attempt to synthesize UTSA-16 on ITO glass and then confirm there is any interaction between them [11]. In our research, we have done one part of the analysis of XRD, UV-Vis absorption and SEM. For the purpose of verifying whether samples have the property of crystallinity, XRD was used to verify the formation of both the TiO_2 and UTSA-16 components in the composites. According to Bragg's law established by Lawrence Bragg and William Henry Bragg in 1913, only crystalline solids generated amazing patterns of reflected X-rays. And at some specific wavelengths and incident angles, there were several intense peaks of reflected radiation [12]. The relation of the wavelength λ and angle is θ

$$2d \, Sin\theta = n\lambda, \quad (1)$$

where n is a positive integer and d is interplanar distance in a crystalline solid. [13] When the result of x-ray diffraction is in line with equation (1), the pattern indicates that sample has great property of crystallinity. Figure 2a and Figure 2b show a XRD pattern of the UTSA-16 crystal on TiO_2 nanoparticles where the volume of ethanol/ deionized water (1/1 v/v) were 10mL/10mL and 7.5mL/7.5mL respectively. As you can see in the two patterns, the main diffraction peaks of Figure 2a are observed at 7.3°, 25.1°and 32.6°.On the other hand, the manifest diffraction peaks of Figure 2b are located at 7.3°, 27.8°and 32.6°. Those peaks present the high crystallinity of the UTSA-16. Obviously, the strongest peak which is at 7.3° is identical to the previous paper [7]. Nevertheless, there are some other peaks appearing in the following figures except for 7.3°. These signals might come from the TiO_2 nanocomposites.

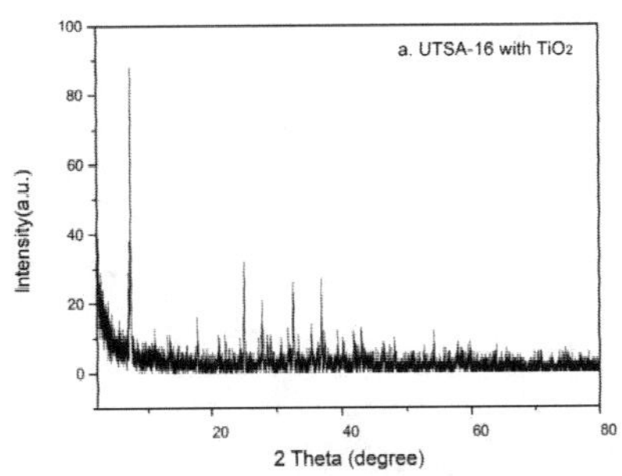

(a)

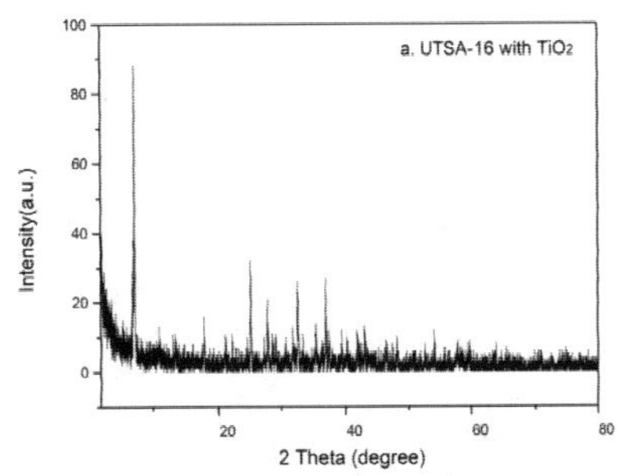

(b)

Figure 2(a and b). XRD pattern of the UTSA-16 crystal on TiO2 nanoparticles (the volume of ethanol/ deionized water (1/1 v/v) were 10mL/10mL and 7.5mL/7.5mL)

By imagining and conceiving the microstructures and morphologies of the composites, SEM brought further evidence of these composites. Figure 3 displays a SEM image of the UTSA-16. These patterns illustrate that the UTSA-16 crystal consists of a large number of irregular particles. These particles look like a variety of tiny pyramids with polyhedral morphology. Figure 4 also shows EDS spectrum of the UTSA-16, which demonstrates there are four elements (cobalt, carbon, oxygen, potassium) in this sample.

IOP Conf. Series: Materials Science and Engineering **720** (2020) 012008 doi:10.1088/1757-899X/720/1/012008

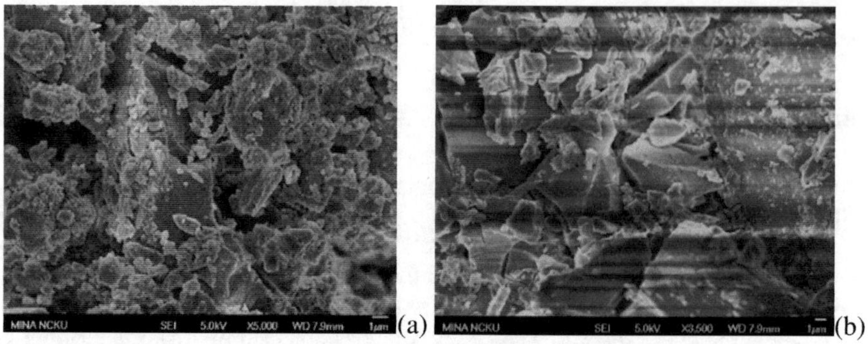

Figure 3. SEM image of UTSA-16

Another analysis is to estimate the UV-Vis absorption. UV/Vis absorption spectroscopy is conventionally used in analytical chemistry for the quantitative determination of different samples, such as transition metal ions, highly conjugated organic compounds and so on. This means that the absorption in the visible light range directly affects the perceived color of the chemical elements involved in the sample. The principal of UV-Vis absorption elaborates that molecules which contain π-electrons or non-bonding electrons can absorb ultraviolet or visible light to excite these electrons to the higher anti-bonding molecular orbitals[14]. In this analysis, we carried out in the presence of solid rather than in solutions. The transition metal ions in solutions can be colored by means of absorbing visible light because d electrons within the metal atoms can be excited from one electronic state to another state. As a consequence, the color of metal ion solutions is strongly affected by the presence of other ligands. As for UTSA-16, the transition metal ion is cobalt ion, and the ligands are water, fully dehydroxylated citrate anion (Cit) and citrate anion (Hcit).

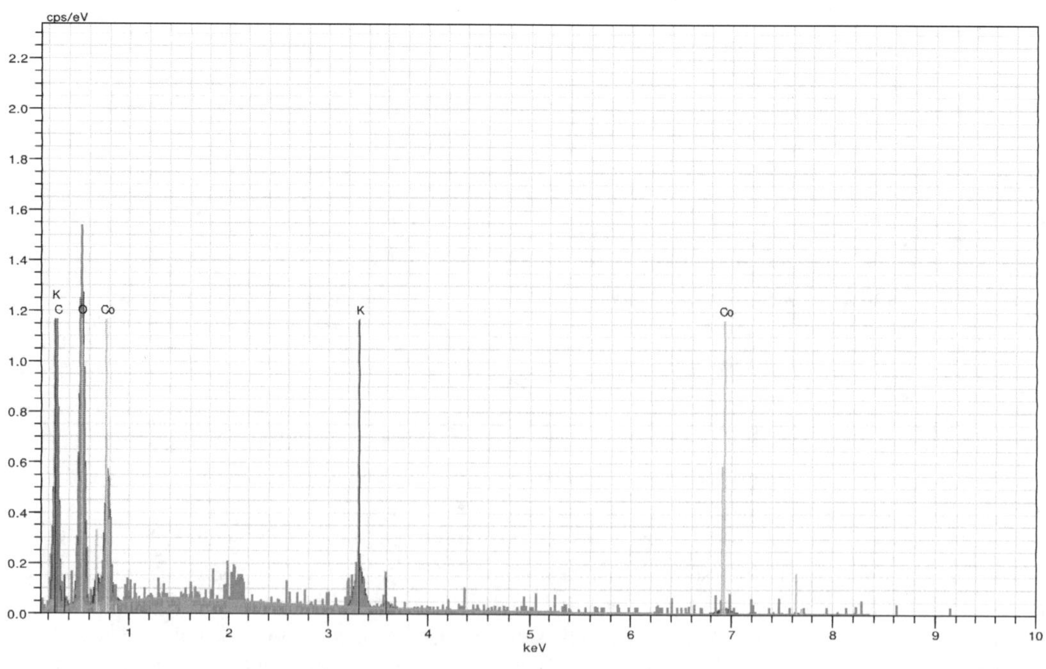

Figure 4. EDS spectrum of the UTSA-16

The statement of Beer's law is that the absorbance of a solution is proportional to the concentration of the absorbing element in the solution[15]. Therefore, for a fixed path length, UV-Vis absorption spectroscopy can be employed to determine the concentration of the absorber in a solution.

In our analysis, the range of light is from 200 to 2500 nm. The UV-Vis absorption spectrum of UTSA-16 on TiO_2 nanoparticles where the volume of ethanol/ deionized water (1/1 v/v) was 10mL/10mL is dominated by two doublets at about 600 and 1200 nm (Figure 5a). And there is another peak at 210 nm in Figure 5a. As for the part of Figure 5b, the UV-Vis absorption spectrum of UTSA-16 on TiO_2 nanoparticles where the volume of ethanol/ deionized water (1/1 v/v) was 7.5mL/7.5mL also have the strongest signals at about 600 and 1200 nm. These two figures are similar to each other except the intensity of absorption. Compared with the UV-Vis absorption spectrum of UTSA-16 in previous study [8], the spectrum was dominated by two doublets at about 500 and 1250 nm. That could be the reason of the difference that TiO2 nanoparticles was on our composites sample.

There is another discussion that the result of UV-Vis absorption spectroscopy is depended by crystallinity of UTSA-16. Focused on the concentration of UTSA-16 which total volume of ethanol/ deionized water is 10mL/10mL, it has stronger intensity of XRD peak at 7.3°. But for the UV-Vis absorption spectrum, the absorption is a bit weaker compared to the other concentration.

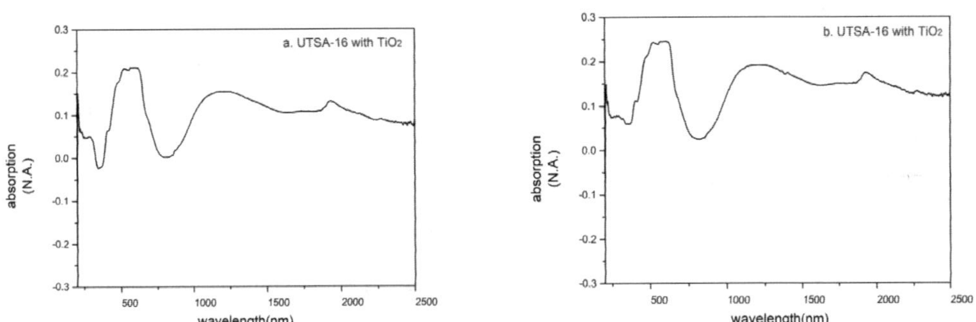

(a) (b)

Figure 5 (a and b). The UV-Vis absorption spectrum of UTSA-16 on TiO_2 nanoparticles (the volume of ethanol/ deionized water (1/1 v/v) were 10mL/10mL and 7.5mL/7.5mL)

4. Conclusion

The leaching efficiency of vanadium is 98.1% at the optimum condition of pH 12.5.In a nut shell, we synthesized the bioinspired cobalt–citrate metal–organic framework (UTSA-16) through a simple solvothermal route on the TiO_2 nanoparticles with different concentration of UTSA-16. However, there are still many unfinished experiments and analysis which now we're trying to work. And we'll check the characteristics of these samples in the future.

Acknowledgments

This work was supported by National Chung Kung University and the Ministry of Science and Technology of Taiwan under 106, Sec. 2, Heping E. Rd., Taipei 10622, Taiwan, ROC, from projects 108-2221-E-006-138, 105-2221-E-006-206 -MY3, 107-2218-E-006-047, and 107-2218-E-006-050.

Furthermore, we express our greatest thanks to Prof. Jyh-Ming Ting, Prof. Jen-Sue Chen, and Prof. Jih-Jen Wu for providing us with advice and discussion.

References

[1] Pachauri, R. K. & Reisinger, A. IPCC Fourth Assessment Report (Intergovernmental Panel on Climate Change, 2007).

[2] Quadrelli, R. & Peterson, S. The energy-climate challenge: recent trends in CO2 emissions from fuel combustion, Energy Policy 35, 5938–5952 (2007).

[3] Yangcan Shen, Ziyin Li, Lihua Wang, Yingxiang Ye, Qing Liu, Xiuling Ma, Qianhuo Chen, Zhangjing Zhang and Shengchang Xiang, Cobalt–citrate framework armored with graphene oxide exhibiting improved thermal stability and selectivity for biogas decarburization, J. Mater. Chem. A, 2015, 3, 593-599

[4] N. MacDowell, N. Florin, A. Buchard, J. Hallett, A. Galindo, G. Jackson, C.S.Adjiman, C.K. Williams, N. Shah, P. Fennell, An overview of CO_2 capture technologies, Energy Environ. Sci. 3 (2010) 1645–1669.

[5] A. Schneemann, V. Bon, I. Schwedler, I. Senkovska, S. Kaskel and R. A. Fischer, Flexible metal–organic frameworks, Chem. Soc. Rev., 2014,43, 6062-6096.

[6] Shengchang Xiang, Yabing He, Zhangjing Zhang, Hui Wu, Wei Zhou, Rajamani Krishna & Banglin Chen, Microporous metal-organic framework with potential for carbon dioxide capture at ambient conditions, Nature Communications volume 3, Article number: 954 (2012)

[7] Jing Jiang, Lan Huang, Xiaomin Liu, and Lunhong Ai, Bioinspired Cobalt−Citrate Metal−Organic Framework as an Efficient Electrocatalyst for Water Oxidation, ACS Appl. Mater. Interfaces, 2017, 9 (8), pp 7193–7201

[8] Masala A1, Vitillo JG, Bonino F, Manzoli M, Grande CA, Bordiga S, New insights into UTSA-16, Phys Chem Chem Phys. 2016 Jan 7;18(1):220-7.

[9] H. Xu, S. Ouyang, L. Liu, P. Reunchan, N. Umezawa, J. Ye, Recent advances in TiO_2-based photocatalysis, J. Mater. Chem. A 2 (2014) 12642–12661.

[10] Li B, Wen HM, Cui Y, Zhou W, Qian G, Chen B, Emerging Multifunctional Metal-Organic Framework Materials, Adv Mater. 2016 Oct;28(40):8819-8860

[11] Angus Crake, Konstantinos C. Christoforidis, Andreas Kafizas, Spyridon Zafeiratos, Camille Petit, CO2 capture and photocatalytic reduction using bifunctional TiO2/MOF nanocomposites under UV–vis irradiation, Applied Catalysis B: Environmental Volume 210, 5 August 2017, Pages 131-140

[12] Bragg, W.H.; Bragg, W.L. (1913). "The Reflexion of X-rays by Crystals". Proc. R. Soc. Lond. A. 88 (605): 428–38. Bibcode:1913 RSPSA.88.428B. doi:10.1098/rspa.1913.0040.

[13] H. P. Myers (2002). Introductory Solid State Physics. Taylor & Francis. ISBN 0-7484-0660-3.

[14] Metha, Akul (13 Dec 2011). "Principle". PharmaXChange.info.

[15] Metha, Akul (22 Apr 2012). "Derivation of Beer-Lambert Law". PharmaXChange.info.

AUTHOR INDEX